ABBREVIATIONS FOR UNITS

A	ampere
amu	atomic mass unit
Å	ångstrom (10^{-10} m)
b	barn (10^{-24} cm)
C	coulomb
°C	degree Celsius
cm	centimeter
eV	electron volt
fm	femtometer, fermi (10^{-15} m)
g	gram
gmw	gram molecular weight
Hz	Herz (s^{-1})
J	Joule
K	kelvin
kg	kilogram
km	kilmeter
keV	thousand electron volts (10^3 eV)
m	meter
MeV	million electron volts (10^6 eV)
mm	millimeter (10^{-3} m)
ms	millisecond (10^{-3} s)
Nt	newton
nm	nanometer (10^{-9} m)
ns	nanosecond (10^{-9} s)
s	second
W	watt
μs	microsecond (10^{-6} s)

$$E_0 = \frac{m_e K^2 e^4}{2 \hbar^2} = 13.6 \text{ eV}.$$

MODERN
PHYSICS
FOR
APPLIED
SCIENCE

MODERN PHYSICS FOR APPLIED SCIENCE

Barry C. Robertson

Queen's University
Kingston, Canada

JOHN WILEY AND SONS
New York Chichester Brisbane Toronto

Library of Congress Cataloging in Publication Data:
Robertson, Barry C   1941-
   Modern physics for applied science.

   Includes indexes.
   1.   Physics.   I.   Title.
QC21.2.R6   530   80-21884
ISBN 0-471-05343-0

Printed in the United States of America
10 9 8 7 6 5 4 3 2 1

PREFACE

It is possible to make two working definitions of modern physics. In the first definition it is any physical property of nature that classical physics (Newton's laws) cannot explain adequately. This is the definition that a physicist would most likely expect to use. In the second definition modern physics is any development that has occurred in approximately the last fifty years that is based on physical principles. This latter definition is quite likely to be regarded as the more reasonable one by people who are only peripherally involved in physics, but in most physics courses it would be considered inappropriate to the main aim in "educating" nonphysics students. While it is true that many developments of great practical impact have arisen from the topics of the first definition, such as transistors, lasers, superconductors and nuclear reactors, there also have been innovations of comparable importance that required only an understanding of classical physics to develop, such as electron microscopy and holography. This kind of development is certain to continue into the foreseeable future. Therefore it seems only reasonable that both definitions should be encompassed in any study of modern physics for people who are inclined to relate what they learn to what they do.

Accordingly this book has the parallel purpose of providing a coherent development of the principles that determine the properties of atoms, molecules, solids, and nuclei, and also of showing the application of physical principles, both classical and modern, to determine ways in which understanding of the cause of various phenomena may be put to practical use. The material assumes a prior introductory course to classical physics, approximately at the level of Resnick and Halliday, and is intended as an introduction to modern physics.

Since the greatest practical potential of physics lies in applications that have not yet been recognized or developed, emphasis in this book has been placed on the *means* by which new uses can be recognized and exploited. For example, whenever a property of an object is influenced by some kind of interaction, the modification of that property can in principle be used as a *measure* of the interaction. Also, similarities between formal properties can be exploited to extend existing knowledge and techniques to new topics. Also, considerable use has been made of mathematical developments that are recognized as only approximately correct in order to show that recognizing and mathematically handling the main significant properties of only partially understood or described phenomena can produce much valid and useful information. Often satisfactory description does not have to await complete comprehension.

In keeping with this approach, the problems at the end of each chapter are not designed to produce familiarity at manipulating "important" formulas. Rather in many cases they are an attempt to encourage a more thorough understanding of the underlying physics by forcing problems to be solved in a slightly unfamiliar "environment." It will be found that some of the problems are rather open-ended, and require considerable thought and work. If presented in the right context these can be quite valuable in showing the students the extreme difficulty and complexity of solution involved in most problems that are of "real" value. It must also be recognized that the best possible problems are ones set up by the instructor of individual classes in response to students' questions and discussion.

The book is not structured in a strongly formal manner. Nor is the use of units. There seem to be several different sets of "natural" units, corresponding to the different regimes that the various topics fall into. The underlying system of units, however, is MKS, and would be SI if that system were not so rigid in disallowing the concurrent use of so many comfortable units.

ACKNOWLEDGMENTS It is a pleasure to recognize that this work has been materially aided by many people. Those that have contributed through countless discussions include Jim Allen, Allan Bridle, Rob Douglas, George Ewan, John Harrison, Dick Henriksen, Hamish Leslie, W.B. Lewis, Paul Lee, Tom McMullen, Mike Sayer, Stan Segel, Malcolm Stott, Bruce Winterbon, Howard Wintle, and Eugene Zaremba. I would also like to thank John Groh, Will Henry, Alec Stewart, and Don Taylor for reading and commenting on the manuscript at various stages of its development. I am especially grateful to Dr. L. Wegmann of Balzers for his very considerate assistance on the subject of photoemission microscopy, Professor L.M. Lidsky of M.I.T. for providing me with information on hybrid fusion devices, and Professor J.D. Sullivan of the University of Illinois for his careful reading of the manuscript and his highly valued comments.

I also thank Alison Cunninghame and Phyllis Bentley for the typing of the innumerable drafts, Tom Leonard for preparing the figures and Lynne Howard for her ceaseless assistance in the search for information.

Barry C. Robertson

CONTENTS

MODERN PHYSICS FOR APPLIED SCIENCE

ONE

SPECIAL RELATIVITY

One of the fundamental relations underlying classical physics is the way that events or observations are transformed from one frame of reference to another. The relation between the description of an event observed from two different frames of reference that are moving at a constant velocity v_0 relative to each other is given by the Galilean transformations, which are

$$x' = x - v_0 t$$
$$y' = y$$
$$z' = z$$
$$t' = t$$

Here the primed frame of reference is taken to be traveling in the $+x$ direction and to have coincided with the unprimed frame of reference at $t = 0$. One consequence of these transformations can be seen by a time differentiation of the equations which gives a relationship between velocities in the two frames of reference;

$$\mathbf{v} = \mathbf{v}' + \mathbf{v}_0$$

This result can be confirmed from direct observation of the behavior of physical objects, and seems perfectly reasonable. However, it must be recognized that our expectation of what is "right," as well as most confirming measurements, are drawn from experience over a limited range of velocities.

1.1 THE MICHELSON-MORLEY EXPERIMENT There is no apparent reason to suspect that the Galilean transformation should not be applicable for velocities

1

approaching that of light. However, around the turn of the century an experiment was set up by Michelson and Morley that effectively tested just this assumption. Michelson and Morley attempted to measure the expected change in the velocity of light when viewed from different frames of reference. Their apparatus was constructed using a light source and a set of mirrors; a simplified setup is shown in Fig. 1-1. Light from the source was split by a half-silvered mirror into two mutually perpendicular beams. These two beams were subsequently added together so that the intensity of the light falling on a screen was determined by the sum of their amplitudes. The whole apparatus could be rotated so that the direction of either of the two arms shown in Fig. 1-1 would allow the light path in that arm to be parallel to the direction of the earth's motion around the sun, v_e. In this way the expected transformation of velocity into a moving frame (parallel to v_e) should lead to a difference in transit times through the parallel and perpendicular (a "stationary" frame) arms, so that the two beams of light will have undergone a different number of cycles when they are recombined, resulting in some amount of cancellation.

The exact amount of cancellation expected depends on the detail of the flight times in the two arms. We will take arm one to be parallel to v_e and arm two to be perpendicular to v_e. For the first arm, the total transit time is given by

$$t_1 = \frac{l}{c - v_e} + \frac{l}{c + v_e} = \frac{l(c - v_e + c + v_e)}{c^2 - v_e^2}$$

For the second arm the velocity is unchanged but the light pathlength is increased by the displacement of the appartus over the transit time interval, as can be seen in Fig. 1-2, so that

$$t_2 = \frac{2\sqrt{l^2 + \frac{v_e^2 t_2^2}{4}}}{c} = \frac{2l}{\sqrt{c^2 - v_e^2}}$$

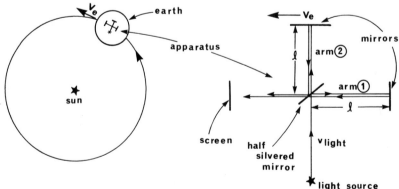

FIGURE 1-1 Simplified arrangement for Michelson-Morley experiment.

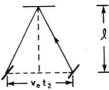

FIGURE 1-2 Path of light in arm two of Michelson-Morley interferometer.

This gives an expected transit time difference of

$$\Delta t = t_1 - t_2 = \frac{2lc}{c^2 - v_e^2} - \frac{2l}{\sqrt{c^2 - v_e^2}}$$

which, after some manipulation and expansion in terms of the small quantity $\frac{v_e}{c}$,

becomes

$$t_1 - t_2 = \frac{lv_e^2}{c^3} + \text{smaller terms}$$

From this we can see that the Galilean transformations definitely show that it should take longer for light to go through arm one than arm two, and that the difference depends explicitly on the earth's velocity. This time difference would be observed as a dependence of the light intensity at the screen upon the orientation of the apparatus with respect to v_e. The maximum intensity should occur when both arms are at 45° to v_e (no transit time difference). The minimum intensity should occur when the transit time difference corresponds to one half a wavelength of the light source. The apparatus angle at which this occurs depends on the value of l. It was not possible to actually observe the point where the full minimum, or complete cancellation of the two beams occurred with the Michelson-Morley apparatus, since the effective value of l (approximately 11m) they used did not produce a sufficiently large transit time difference. Nevertheless the measurement of the maximum light intensity reduction (achieved when one or the other of the two arms of the apparatus is parallel to v_e) could be used to test the value predicted using the Galilean transformations. In the actual experiment performed by Michelson and Morley the sensitivity of the measurement was doubled by observing the intensity change on rotating the apparatus through a 90° arc from a position where arm one, for example, was parallel to v_e to a position where it was perpendicular to v_e.

If the known orbital velocity of the earth is inserted into the time difference equation together with the 11m effective value for l used by Michelson and Morley, the calculated transit time difference from the above equation corresponds to approximately 0.2 wavelengths of the light source. Consequently when the apparatus was rotated through the 90° arc the relative phase of the two light beams being added together was expected to change by approximately 0.4 wavelengths, which was a large value, compared with the smallest change they could detect, which was approximately 0.01

wavelengths. When the actual experiment was performed *no* measurable change was found. At first it was thought there may have been some additional absolute veolcity in the sun-earth system which just happened to cancel the earth's velocity when the experiment was performed. However the measurement has been repeated since at various times of the day and on various days of the year, and in every case the original null result has been confirmed. There is only one possible conclusion from the experiment—that the measured velocity of light is *completely independent* of the frame of reference of the measurement.*

The constancy of the measured velocity of light ($v \equiv c = 3 \times 10^{10}$ cm/s), although a seemingly simple observation, forces on us a reconsideration of the validity of the Galilean transformation and the associated notion of addition of velocity. That a reevaluation is required can be recognized by considering, for example, the case of measuring the velocity of light emitted from a source that is moving toward us with a velocity of, say, $c/2$. The Michelson-Morley experiment tells us that we will measure the velocity of that light to be c, *not* $c + c/2$, as would be expected from the Galilean addition of velocities.

1.2 CONSEQUENCES OF THE CONSTANCY OF C Conceptually it is very difficult to see how to reconcile the experimental observation that the velocity of light is always measured to be the same value, regardless of the measurer's frame of reference, with what we expected to be true. However there are some straightforward mathematical consequences of such an observation. These can be determined by considering what must be observed in two frames of reference that are moving with respect to one another when a burst of light is emitted from a flashbulb. For mathematical convenience we will assume the flashbulb is at the center of a frame of reference (the primed frame) which is moving with a constant velocity v along the $+x$ direction of a second coordinate system (the unprimed frame). In addition, let the two frames of reference coincide at the instant the flashbulb goes off, which we will choose to be $t = 0$. After the flash a hemispherical shell of light will be observed to travel out uniformly in all directions with a velocity c in *both* frames of reference (Fig. 1-3). The radius of this hemispherical shell is described at any time by

$$x^2 + y^2 + z^2 = c^2 t^2$$

in the unprimed system and

$$x'^2 + y'^2 + z'^2 = c^2 t'^2$$

in the primed system (time in the primed system has been denoted here as t'). It must be possible to transform the observation from one frame of reference to the other, that is, apply a transformation to the radius equation in the primed system and get the corresponding equation in the unprimed system, since both equations correctly de-

* For a detailed discussion of this experiment see Further Readings, French, Ch. 1.

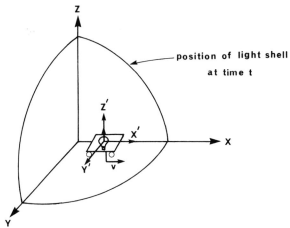

FIGURE 1-3 Position of light flash a time t after it is emitted at the origin from a source moving with a constant velocity v.

scribe the same event. It is easy to show that the Galilean transformation will *not* do this (unless $t = 0$ or $v = 0$) so we must generate a new set of transformations that are capable of correctly describing events when the velocity of light is involved as well as events that involve much lower velocities. Whatever these new transformations turn out to be they must be consistent with the Galilean transformations for low velocities. The easiest way to guarantee this is simply to apply different "correction factors" to the Galilean transformations and find out what ones are necessary to obtain consistency between the two observations in the flashbulb experiment.

The set of transformations

$$x' = \gamma(x - vt)$$
$$y' = y$$
$$z' = z$$
$$t' = ax + bt$$

allows us to modify the behavior of the transformation directly involving v and the one involving t. There is no *a priori* reason for choosing this particular modification, nor do we know what the form of γ, a, and b must be. However we can determine whether or not these modifications can result in consistency, and, if so, what the values of γ, a, and b must be to obtain it, by setting up the equality

$$x^2 + y^2 + z^2 - c^2t^2 = x'^2 + y'^2 + z'^2 - c^2t'^2$$

and use our guessed-at relationships to transform the observation in the primed frame into one in the unprimed frame. We then get

$$x^2 + y^2 + z^2 - c^2t^2 = \gamma^2 (x-vt)^2 + y^2 + z^2 - c^2(ax + bt)^2$$

Since this equality must hold for all values of x, y, z, and t because they are *independent* variables, the coefficients of each of the variables must be the same on both sides of the equality sign. This leads to three required relations

$$\gamma^2 - a^2c^2 = 1 \qquad \text{(from } x\text{)}$$

$$b^2 - \gamma \left(\frac{v}{c}\right)^2 = 1 \qquad \text{(from } t\text{)}$$

$$\gamma^2 v + a\, bc^2 = 0 \qquad \text{(from cross terms)}$$

These are three equations in three unknowns, and we obtain after appropriate manipulation

$$a = \frac{-v/c^2}{\sqrt{1 - (v/c)^2}}$$

$$b = \frac{1}{\sqrt{1 - (v/c)^2}}$$

$$\gamma = \frac{1}{\sqrt{1 - (v/c)^2}}$$

The fact that we are able to obtain a solution for a, b, and γ means that a set of transformation *can* be set up which converts the observation of the light sphere in one frame correctly into the observation made in the other frame. They are

$$x' = \frac{x - vt}{\sqrt{1 - (v/c)^2}}$$

$$y' = y$$

$$z' = z$$

$$t' = t - \frac{vx/c^2}{\sqrt{1 - (v/c)^2}}$$

and are called the Lorentz transformations. They are the simplest ones known that are consistent with the experimental fact that the measured velocity of light is independent of the observer's frame of reference.

These relations have been derived only for constant relative velocities, and their consequences constitute the subject of *special* relativity. How different are the new relations from the old ones? The relation between x and x' has been modified by the factor $1/\sqrt{1 - (v/c)^2}$. For the velocities that we commonly experience, v up to $\sim 10^2$ m/s, the term $(v/c)^2$ is less than $\sim 10^{-12}$; the difference between $1/\sqrt{1 - (v/c)^2}$ and 1 is then negligible. It is a significant correction only when the velocity v is an appreciable fraction of c, say 10 percent or more. Similarly time is the same in two frames of reference except for very large relative velocities or for very large separations. The Galilean transformations are a ''low velocity'' approximation of the Lorentz transformations, and we couldn't have expected to be aware of this until we looked at ''high velocity'' events.

1.3 RELATIVISTIC RULERS AND CLOCKS   Although the fact that the Lorentz transformations become indistinguishable from Galilean transformations for the velocities and distances common to everyday experience, the underlying concepts of space and time have been modified in some way by these new relationships. In order to obtain an idea of what effect these modifications have on the concept of space and time, we can perform simple thought experiments.

First we will examine the effect on space, as seen in the properties of a ruler when viewed from different frames of relative motion with respect to our own. This ruler we will call a test ruler, and its length will be measured against a ''master ruler'' that is kept stationary with respect to ourselves. The length of the test ruler while at rest can be measured by noting its start and end positions when placed against the master ruler; the difference of these two values will be defined as the *rest length, l_0*.

We can repeat this length measurement of the test ruler when it is moving past us (and the master ruler) with a constant velocity v. This can be done in the same manner as before, by marking on the master ruler the positions x_1 and x_2 of the start and end of the moving test ruler at some instant, as illustrated in Fig. 1-4. The length of the moving ruler, which we will denote l, is equal to $x_2 - x_1$; these two position measurements can be transformed into the moving frame of reference according to the Lorentz relations so that

$$x_1' = \frac{x_1 - vt}{\sqrt{1 - (v/c)^2}}, \qquad x_2' = \frac{x_2 - vt}{\sqrt{1 - (v/c)^2}}$$

The way we have fixed x_1 and x_2 means that the value of t is the same in both cases so we obtain the relation

$$x_2' - x_1' = \frac{x_2 - x_1}{\sqrt{1 - (v/c)^2}}$$

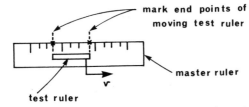

mark end points of
moving test ruler

master ruler

test ruler

v

FIGURE 1-4 Measurement of length of a moving ruler.

The interval $x_2 - x_1$ is the length l we measured for the moving ruler, and the interval $x_2' - x_1'$ is the length l_0 of the ruler as measured in its own frame of reference.

This gives us the relation

$$l_0 = \frac{l}{\sqrt{1 - (v/c)^2}}$$

or

$$l = l_0 \sqrt{1 - (v/c)^2}$$

The existence of the Lorentz transformations requires that the measured length of the test ruler depends on its velocity relative to the measurer. In general the length of an object must decrease with increasing relative velocity. This effect is called the Fitzgerald contraction and, although it is not normally a large effect, it can be important. It is this contraction of length in association with charge density in currents which is the basic source of the magnetic force.*

An even simpler situation will reveal the effect of relative motion on the concept of time. We will construct a clock that consists of a light beam that is reflected between two parallel mirrors placed a distance L apart. The period of the clock when observed in its own frame of reference can be defined as T_0, where

$$L = \frac{c T_0}{2}$$

The period of the clock can be remeasured when it is set in motion with a relative constant velocity v and is oriented as shown in Fig. 1-5 so that the light path is perpendicular to the direction of motion of the clock. Because of the motion of the clock, the light beam no longer goes straight up and down, but takes the path illustrated in Fig. 1-5. The observed velocity of light in the moving clock must still be c, and we find that the period of the moving clock, T, is fixed by the "geometry" of the light path so that

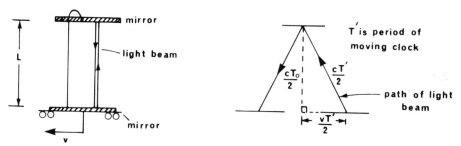

FIGURE 1-5 Measurement of time in a moving frame of reference using a light beam.

* For a treatment of the relativistic origin of the magnetic force, see Further Reading, Sears and Zemansky, sec. 30-2.

$$\left(\frac{v\,T}{2}\right)^{2} + \left(\frac{c\,T_0}{2}\right)^{2} = \left(\frac{c\,T}{2}\right)^{2}$$

which leads to the relation

$$T = \frac{T_0}{\sqrt{1 - (v/c)^2}}$$

The measured period of the moving clock cannot be the same as the measured period of that same clock when it is stationary. Clearly, the *reason* is because we have required that the velocity of light is c, regardless of whether the source is moving or not. We can see qualitatively from the figure that the measured period of the clock has to be longer when it is moving than when it is not moving, and the formula we have just obtained confirms this. This effect is called *time dilatation*.

There is a strong temptation to ascribe these effects (contraction of length and time dilatation) simply to mathematical artifacts which are not connected with reality. This is not so. They have been obtained as logical consequences of the *observed fact* that the measured velocity of light does not change with an observer's frame of reference. Furthermore there are several direct experimental verifications of these effects. Perhaps the most straightforward experimental verification has been demonstrated with the use of elementary particles called π-mesons. They are particles that are unstable and that spontaneously decay into other elementary particles in such a way that the time of decay can be fixed experimentally. Their decay can be characterized by a mean lifetime, denoted τ, which has been measured to be $\tau = 2.6 \times 10^{-8}$ s (in the particles' own frame of reference). Pi-mesons can also be produced readily using high-energy particle accelerators in such a way that they travel with a constant, well-defined high velocity until they decay (Fig. 1-6). Since we can know how fast they are traveling and where they start, we can calculate the mean distance they will travel before they decay, and then test the calculation experimentally. In an actual experiment performed for this purpose, the π-mesons were produced with a velocity $v = 0.75\,c$, and would therefore be expected to travel a mean distance

$$l = v\tau = 0.75 \times 3 \times 10^8 \text{ m/s} \times 2.6 \times 10^{-8} \text{ s} = 5.9 \text{ m}$$

The experimentally measured mean distance was $l = 8.5 \pm 0.6$ m, which is in significant disagreement with our calculated value. However, if we recognize that the π-mesons are effectively moving clocks, we realize that we have to transform their measured time (the π-meson lifetime) into our frame of reference in order to deal with them correctly. Therefore their mean lifetime when moving will be given by

$$\tau_{\text{moving}} = \frac{\tau}{\sqrt{1 - (v/c)^2}}$$

$$= \frac{\tau}{\sqrt{1 - (0.75)^2}} = 4 \times 10^{-8} \text{ s}$$

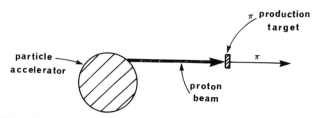

FIGURE 1-6 Production of *pi* mesons which are traveling at a high velocity.

and the mean distance they should travel is actually

$$l = 0.75 \times 3 \times 10^8 \text{ m/s} \times 4 \times 10^{-8} \text{ s} = 8.9 \text{ m}$$

which is in agreement with the measured value, within the errors of the measurement. This is a *direct confirmation* of a relativistic effect. Special relativity is not just a mathematical quirk, but a physical reality.

1.4 RELATIVISTIC MASS Our previously inadequate conception of space and time must lead to significant alterations of the details of dynamics if we are to have a correct description of objects at all speeds. Moreover, we should expect to find it necessary to alter our understanding of mass, since we have defined this concept effectively through the dynamic behavior of entities with mass.

In order to see most easily what the effect of relativity will have on mass, we will make a measurement of mass through a collision experiment, first in a single frame of reference, and then in two frames of reference moving with respect to each other. We will do the measurements by means of two observers who each have a ball whose mass, when determined in the same frame of reference, is exactly the same. This can be confirmed by throwing the balls directly at each other at the same speed and in such a way that the balls collide midway between the two observers. (These conditions are not necessary, but only serve to make the calculations simpler.) If the observers measure the speed of the balls as they bounce directly back, the relative masses can be determined by using the law of conservation of momentum. This experiment can be repeated when the observers are in different frames of reference which have a constant relative velocity v.

Each observer throws his ball out perpendicular to the direction of relative motion in such a way that the two balls collide midway between the observers and bounce directly back.

We can view this from several different frames of reference; if the frame of each observer is given equal and opposite velocity $v/2$, we see the situation as shown in Fig. 1-7a.

The behavior of one and two is completely symmetric. For identification purposes we have labeled one's ball M_A and two's ball M_B. If we look at this experiment from

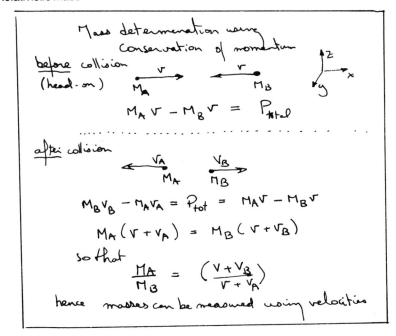

Mass determination using
conservation of momentum

before collision
(head-on)

$M_A v - M_B v = P_{total}$

after collision

$M_B V_B - M_A V_A = P_{tot} = M_A v - M_B v$

$M_A (v + V_A) = M_B (v + V_B)$

so that

$$\frac{M_A}{M_B} = \left(\frac{v + V_B}{v + V_A} \right)$$

hence masses can be measured using velocities

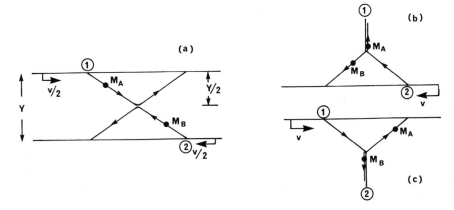

FIGURE 1-7 Collision experiment observed from three different frames of reference; (a) where one and two are traveling with equal and opposite velocities, (b) where one is at rest and (c) where two is at rest.

one's point of view we see events as they appear in Fig. 1-7b. Observer one sees his ball go straight out and straight back, traveling a total distance Y in a time we will call T_0 so he knows the ball had a velocity $(v_A)_1$ according to himself, which is equal to Y/T_0. Similarly, because the situation is completely symmetric, observer two sees events as shown in Fig. 1-7c and he too knows that $(v_B)_2 = Y/T_0$. Now let us go back to observer one; he sees a collision to which he knows he can apply the law of conservation of momentum. Since he threw the ball straight out, he knows that

$$(M_A \ v_A)_1 = (M_B \ v_B)_1$$

where v_B is the component of the velocity of B (as seen by one) that is parallel to $(v_A)_1$. Now he can measure that velocity, and finds it to be

$$(v_B)_1 = \frac{Y}{T}$$

where T is the total transit time of ball B according to him. But that time T is related to the time T_0 that two measured for the same event by the relation

$$T = \frac{T_0}{\sqrt{1 - (v/c)^2}}$$

(note that Y is the same for both observers).

Therefore observer one concludes that

$$(M_A v_A)_1 = (M_B v_B)_1$$

$$M_A \frac{Y}{T_0} = M_B \frac{Y}{T}$$

$$= M_B \frac{Y}{T_0} \sqrt{1 - (v/c)^2}$$

$$M_A = M_B \sqrt{1 - (v/c)^2}$$

The two balls do *not* have the same mass now. Furthermore, the choice to follow observer one's calculations was completely arbitrary; we could as easily have followed two's calculation, and would have found that two came to the same conclusion as one, except the mass labels A and B would have been interchanged. What they would agree on, and what in fact is common, is that if we replace A and B with "moving" and "rest" the two results agree. That is

$$M_{\text{moving}} = \frac{M_{\text{rest}}}{\sqrt{1 - (v/c)^2}}$$

Both observers agree that moving objects become more massive.

This apparently self-contradictory situation has been observed experimentally. In particular there are several different machines for accelerating elementary particles to very high energies which will not operate properly unless the relativistic variation of mass is accounted for. These machines are based on the principles of a cyclotron,

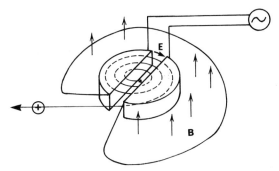

FIGURE 1-8 Schematic diagram of cyclotron. The alternating electric field E is in the direction of the charged particles' motion, and the magnetic field B is transverse to their direction of motion.

which is outlined schematically in Fig. 1-8. Charged particles are produced at the center of the cyclotron, where they experience an acceleration due to the presence of an electric field set up by the hollow D-shaped electrodes. The particles are accelerated into the hollow electrode where they are shielded from the electric field, but still experience an acceleration at right angles to their motion due to the presence of a uniform magnetic field in which the whole device is immersed. This causes the particles to circle back to the gap between the electrodes (called *dees*). If the polarity of the electric field is reversed at this time, the particles will experience a further acceleration on crossing the gap between the dees, circle around again, this time in a larger orbit, and approach the gap once again from the original direction. From the uniform circular motion of free particles in a uniform magnetic field we know that particle orbits have a frequency (called the cyclotron frequency) of

$$\nu_{\text{cyc}} = \frac{qB}{2\pi m}$$

where B is the value of the magnetic field and q and m are the particles' charge and mass, respectively. If the electric field produced in the gap between the dees is made to oscillate with the same frequency, particles that are accelerated in one cycle will continue to be accelerated to higher and higher energies in subsequent cycles. Since the cyclotron frequency of the particles depends only on B (a constant) and the fundamental properties of the particles, charge and mass, the frequency at which the electric field should be alternated should be fixed simply by the type of particle being accelerated and be completely independent of energy.

It is found, however, that as the particles approach extremely high energies their angular velocity begins to slow down, so that they lag behind the variation of the electric field and cease to be accelerated any more. The reason is that their mass has increased due to their high speed, and any further acceleration can only be achieved by continuously matching the frequency of the electric field oscillations to the particles' cyclotron frequency as fixed by their current velocity and mass. Cyclotrons designed to do this job are called synchro-cyclotrons. Their design explicitly relies on the relativistic variation of mass we have arrived at.

1.5 THE ULTIMATE SPEED AND MASS-ENERGY Relativity requires a complete, detailed reevaluation of all dynamics; however we can retain Newtonian mechanics both as a low-velocity approximation and as a starting point to find out what kinds of important changes to expect. The basic working formula for physics and engineering has been Newton's second law which we effectively used as a prescription to deduce an object's motion from the forces acting on it, that is,

$$\mathbf{a} = \frac{\mathbf{F}}{m}$$

Now we know that this is only approximately correct; not only does a force cause an acceleration of the object but it also brings about an increase in the object's mass as well. If we assign a zero-motion mass, or *rest mass* to the object of m_0 then the acceleration of the object is given approximately by

$$\mathbf{a} = \frac{\mathbf{F}}{m_0} \sqrt{1 - \left(\frac{v}{c}\right)^2}$$

From this we can see that for any finite force applied to a massive object the acceleration produced must approach zero as the velocity of the object approaches c. It is therefore not possible to accelerate any physical object to the speed of light. Effectively what happens is that the object would become infinitely massive, requiring an infinite amount of work to be done on it to attain the speed of light.

There is one more aspect of relativity of interest to us, which is hinted at by thinking about what is happening in the above example of attempting to accelerate an object to very high speeds. However, we can approach the problem with more mathematical directness by returning to the formula for a moving mass

$$m = m_0 \left[1 - \left(\frac{v}{c}\right)^2\right]^{-1/2}$$

where again m is the mass while in motion and m_0 is the rest mass. Since (v/c) is normally a very small quantity we can make use of the approximation

$$(1 + \delta)^N \cong 1 + N\delta$$

where δ is a small number compared to 1. This means that we can approximate the mass relation by.

$$m \cong m_0 \left[1 + \tfrac{1}{2} \left(\frac{v}{c}\right)^2\right]$$

$$= m_0 + \frac{m_0 v^2}{2 c^2}$$

or

$$mc^2 = m_0 c^2 + \frac{m_0 \, v^2}{2}$$

Looking at this equation we recognize the last term as kinetic energy. Any mathematically correct relation must have the same units for all terms; therefore mc^2 and $m_0 c^2$ must also have units of energy and must describe some form of energy. In fact we have arrived at Einsten's famous relation

$$E = mc^2$$

which explicitly states the equivalence of mass and energy and implies that they are interconvertible. In fact the conversion of energy (in the form of work) to mass has already been encountered, both in discussing what happens as an object approaches the velocity of light and in the "practical" case of the synchro-cyclotron where the energy in the oscillating electric field was used to increase both the particle's velocity and its mass.

Mass and energy can no longer be regarded as separate entities, obeying separate conservation laws. Rather, a single mass-energy law holds, where these two entities are interchangeable. A general relation between mass, energy, and momentum can be obtained which, in a way, is analogous to the more familiar relationship between total, kinetic, and potential energy. We know that the momentum p is given by

$$p = mv = \frac{m_0 v}{\sqrt{1 - (v/c)^2}}$$

$$p = mr = \frac{m_0 v}{\sqrt{1 - (v/c)^2}}$$

and similarly

$$E = mc^2 = \frac{m_0 c^2}{\sqrt{1 - (v/c)^2}}$$

By taking the difference of the squares of these two relations we can obtain

$$E^2 - c^2 p^2 = m_0^2 c^4$$

or, by rearranging and taking the square root of both sides,

$$E = \sqrt{c^2 p^2 + m_0^2 c^4}$$

This relation gives the familiar expression for kinetic energy, $p^2/2m_0$, as the leading term of an expansion of $E - m_0 c^2$ when p is much smaller than $m_0 c^2$.

FURTHER READING   M. Berry, *Principles of Cosmology and Gravitation,* Cambridge University Press, 1976.

M. Born, *Einstein's Theory of Relativity,* Dover, 1965.

A. P. French, *Special Relativity,* Norton, 1968.

J. Norwood, Jr., *Twentieth Century Physics,* Chapter 3, Prentice-Hall, 1976.

W. Rindler, *Essential Relativity: Special, General and Cosmological,* Van Nostrand–Reinhold, 1969.

F. W. Sears and M. W. Zemansky, *University Physics,* 3rd edition, Section 30-2 (magnetism), Addison–Wesley, 1964.

PROBLEMS   1. Supersonic jets can travel at speeds up to $\sim$ 1500 mi/hr. How accurate would a clock on board have to be if relativistic effects due to the speed were to contribute significantly (say 10%) to the sources of the clock's inaccuracies? Is any clock capable of this accuracy?

2. Prove that the transit time difference in the two arms of the Michelson-Morley apparatus is given by $t_1 - t_2 \cong lv_e^2/c^3$. What is the fractional error inherent in this approximation?

3. Confirm that a $\sim 0.4\lambda$ shift would be expected for the Michelson-Morley apparatus due to the motion of the earth around the sun. What is a reasonable value to use for λ?

4. Describe the π meson experiment in terms of what the π meson "sees." Is this compatible with what is observed in the laboratory frame? Explain.

5. At what speed will a cube become a block whose thickness in the direction of motion is half its width?

6. According to an observer on earth a light pulse starting from the earth reaches the moon in 1.21 s. How long will this pulse take to reach the moon, according to an observer moving in the same direction as the pulse with a speed (a) 0.1 c, (b) 0.5 c, (c) 0.9 c?

7. A large nucleus is fired from a nuclear accelerator with a velocity $v = 0.75$ c. In its own frame of reference the projectile is elongated into a cigar shape whose major axis is twice as large as its minor axis, and oriented at 45° to the direction of motion. What is the shape of the projectile in the laboratory frame of motion?

8. Using the Lorentz transformations show that an object moving with velocity components u'_x, u'_y in a primed system, which itself has a relative motion v in the x direction, will have a motion in the unprimed system described by

$$u_x = \frac{u'_x + v}{[1 + (u'_x \, v/c^2)]}, \quad u_y = \frac{u'_y}{\gamma[1 + (u'_x \, v/c^2)]}$$

9. Two beams of elementary particles are moving toward each other, each with a velocity $v = 0.6$ c in the laboratory frame of reference. What is their velocity of approach with respect to each other?

10. A heavy nucleus is fired from a nuclear accelerator with a velocity $v = 0.35$ c at a target that has half the rest mass of the projectile. The two nuclei fuse, then, in the center of mass frame, break up into two fragments of equal mass which are emitted perpendicular to the incident direction. Assuming there is no change of internal energy, at what lab angle are the fragments emitted?

11. Through what potential must an electron fall (ignoring relativity) in order to acquire a velocity equal to the speed of light? What speed does the electron actually acquire in falling through this potential?

12. An object has a rest mass of 2×10^{28} kg. What does this correspond to in eV?

13. Show that $(1 + A)^n \cong 1 + nA$ for small A. Estimate the error from using this approximation.

14. The energy released in the fission of a ^{235}U nucleus is approximately 200 MeV. What fraction of the nucleus mass is converted to energy?

15. A beam of electrons is accelerated through a 750 kV potential difference. If they start from rest, what is their final velocity? What would their final velocity be if they started with $v/c = 0.1$?

16. Using the fact that $F = d(mv)/dt$ show that the work W done on a particle of rest mass m_0 to accelerate it to a speed v is

$$W = m_0 c^2 \left(\frac{1}{\sqrt{1 - (v/c)^2}} - 1 \right) = KE$$

What percentage error is made in using $KE = m_0 v^2/2$ for a particle if its speed is (a) 0.01 c, (b) 0.1 c, (c) 0.5 c?

17. Show that the relation $E^2 = p^2 + m^2 c^4$ is correct.

18. A particle with rest mass m_0 approaches another particle with rest mass m_0. Each is moving at a speed of 0.75 c. Calculate the relativistic mass of each particle with respect to the other.

TWO

THE ELECTRON

In a practical sense modern physics started with the discovery of electrons and their potential usefulness. They have been and continue to be of great use in many devices, and their study also leads in a natural way to the development of many of the new concepts in modern physics.

The properties of the electron are described most basically by these facts:

$$q_e = -1.602 \times 10^{-19} \text{ C}$$

$$m_e = 9.110 \times 10^{-31} \text{ kg}$$

For comparison the electron's charge is equal in magnitude to that of the proton but its mass is only about 1/2000 of the proton mass.

From the electron's properties, charge and mass, we must first decide which of the two associated forces, Coulomb and gravitational, will control its behavior. A direct comparison of the two types of force between two electrons yield

$$\frac{F_c}{F_g} = \frac{kq_1q_2}{Gm_1m_2} \cong \frac{9.0 \times 10^9 \times (1.6 \times 10^{-19})^2}{6.7 \times 10^{-11} (9.1 \times 10^{-31})^2}$$

$$\cong 5 \times 10^{42}$$

where $k(=1/4\pi\epsilon_0)$ is the Coulomb force constant and G is the gravitational constant.

Since both the Coulomb and gravitational forces vary as $1/r^2$, the ratio of their two magnitudes is independent of distance and indicates that the Coulomb force is much greater than the gravitational force. Therefore in all calculations concerning electrons we will pay attention to the Coulomb force (plus its relativistic "correction," the magnetic force) alone. Everything that can be known about the electron's motion is given when we write

19

$$F = q(\mathbf{E} + \mathbf{v} \times \mathbf{B})$$

where $\mathbf{E}$ is the electric field and $\mathbf{B}$ the magnetic field that exists wherever the electron passes with a velocity $\mathbf{v}$.

2.1 FREE ELECTRON EXTRACTION If we want to obtain electrons in order to do something with them, where can we find them in a readily available form? They exist in all materials, of course, but in many they are very close to, and therefore strongly bound to, the protons that make up the material. However, we know from elementary electricity that the class of materials called conductors have electrons that are free to move anywhere within the conductor. It would seem reasonable to start by trying to extract electrons from conductors.

The idea of electrons freely moving about inside a conductor almost suggests that we ought to be able to remove them simply by tipping the conductor so that they all fall out the bottom. Clearly this is not so. If we think about it, we easily can get a rough picture of why and also of what we might have to do to extract them.

As illustrated in Fig. 2-1, in the interior of a metal the electrons experience forces due to the protons which cancel each other so that their motion is virtually unrestricted. However this is no longer true at the surface.

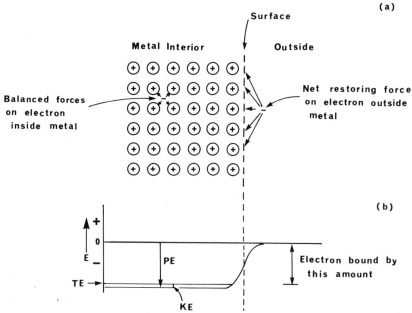

FIGURE 2-1 Forces on electrons inside and at the surface of a metal. An electron with a kinetic energy *KE* inside the metal is bound into the metal by the potential energy *PE*, which represents the amount of work required to remove it from the surface. The total energy of the electron is represented by *TE*.

As the electron moves towards the outside of the surface, a strong unbalanced force on the electron arises which keeps it from escaping. Stated in terms of energy, the electron's kinetic energy *KE* is much smaller than its (negative) potential energy *PE* inside the metal. (Note that we have set the potential energy of the electrons to zero when they are infinitely far away from the protons.)

The electrons are bound into the metal by an amount of energy ϕ (called the *work function*). The actual value of ϕ depends in detail on the nature of the material. Typical values of ϕ range from ~1 to 4 *eV* with most metals closer to 4 *eV*; later we will see how these values arise in detail. At the moment all we need to know is that to remove electrons from the interior of a material we need to do an amount of work on them equal to ϕ per electron. The measured work functions for several materials are listed in Table 2-1.

TABLE 2-1   Work Functions of Typical Metals

Metal	$\phi(eV)$	Metal	$\phi(eV)$	Metal	$\phi(eV)$
Li	2.38	Ca	2.80	In	3.8
Na	2.35	Sr	2.35	Ga	3.96
K	2.22	Ba	2.49	Tl	3.7
Rb	2.16	Nb	3.99	Sn	4.38
Cs	1.81	Fe	4.31	Pb	4.0
Cu	4.4	Mn	3.83	Bi	4.4
Ag	4.3	Zn	4.24	Sb	4.08
Au	4.3	Cd	4.1	W	4.5
Be	3.92	Hg	4.52		
Mg	3.64	Al	4.25		

From *Solid State Physics* by Neil W. Ashcroft and N. David Merwin. Copyright © 1976 by Holt, Reinhart and Winston. Reprinted by permission of Holt, Reinhart and Winston.

2.1a Cold Cathode Emission   How can we actually remove the electrons? The simplest way to think of is to apply an electric field to the electrons in the surface region to balance the forces from the protons. The potential difference required is only a few volts, but it must occur over a distance of a few atomic spacings (~ 10^{-8} cm). Consequently we can expect to require a field on the order of ~ 10^8 V/cm; this is extremely high, but not impossibly so. The process, when it happens, is called cold cathode emission. Because the high fields necessary for cold cathode emission are obtained most readily for regions that have a very small radius of curvature (remember $E = kq/r^2$), this effect has been used to develop a kind of microscope for "seeing" very fine structure down to the size of large atoms. In its simplest form, the so-called field emission microscope consists of the object with the structure to be magnified at the center of an evacuated spherical glass bulb that has been coated with a fluorescent material. A potential difference between the screen and the object then creates the high fields at the points of small radius of curvature. Electrons from these points are ejected radially out to the screen, and give an effective magnification of the surface structure in the ratio of the bulb radius to the point radius which can be of the order of several hundreds of thousands. Systems currently used, however, make use of the ionization of an inert gas such as helium or neon by the same high field regions to form the image, rather than field emission electrons. Operated in this mode, they are called field ion microscopes and an example of the result achievable is shown in Fig. 2-2.

2.1b Thermionic Emission The most common way to extract electrons from a metal, however, is simply to apply thermal energy and "boil" them out. If the electrons can be thought as a free "gas" inside the metal (this isn't quite true, as will be seen later), then it is possible that through collisions with other electrons and the metal atoms some of them will acquire enough kinetic energy to surmount the potential barrier at the metal surface and escape.

The proper description of the energy distribution of a gas is given by the Maxwell-Boltzmann distribution

$$N(E) = \frac{2N_0}{\sqrt{\pi} \, (kT)^{3/2}} \, E^{1/2} \, e^{-E/kT}$$

where $N(E) \, dE$ is the number of particles having a kinetic energy between E and $E + dE$, N_0 is the total number of particles, T is the temperature (degrees Kelvin) and k is the Boltzmann constant. This distribution peaks at $kT/2$ and has the general shape shown in Fig. 2-3.

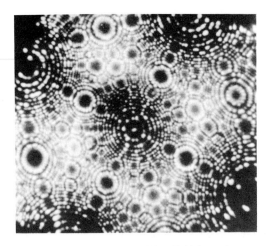

FIGURE 2-2 Gold needle point photographed with a field ion microscope. The approximate tip radius is 250 Å. From R.S. Averback and D.N. Seidman, *Surface Science* 40, 249–263 (1973). Original photograph provided courtesy of D.N. Seidman.

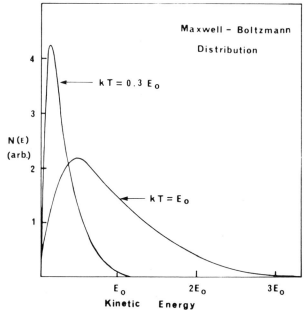

FIGURE 2-3 Maxwell-Boltzman distribution for two values of kT.

At high E compared to kT the distribution is dominated by the exponential term. Clearly the temperature of the metal is critical to the question of whether or not electrons will be emitted, so we need to know what kT is compared to the work function ϕ. For room temperature ($\sim$ 300 K) kT is 1.38×10^{-23} J/K $\times$ 300 or 4.15×10^{-23} J. We can convert this to a more useful quantity by remembering that 1 Coul. Volt = 1 J, so that $kT = 0.026$ eV. Comparing this with ϕ, it is clear from Fig. 2-1 that there is virtually no difference between the work function ϕ and the total potential energy of the electron gas in the metal. This means that we can estimate quite easily the fraction of electrons that have enough kinetic energy to escape if they approach the edge of the metal by finding out the fraction of electrons that have a kinetic energy at least equal to ϕ. This is given by*

$$N(\phi) = N_0 \left(\frac{2(\phi)^{1/2}}{\sqrt{\pi}(kT)^{3/2}} \right) \left(e^{-\phi/kT} \right)$$

By considering the relative magnitude of ϕ, a few eV, and kT, a few hundredths of an eV, we can see that the probability is almost entirely controlled by the exponential term in the equation. Simply from evaluating that term using convenient numbers like $\phi = 2.6$ eV and $kT = 0.026$ eV we find

---

* This formula is not exactly correct for electrons in a metal, but we will see later that the necessary modifications will not change any of our conclusions.

$$N(\phi) \gtrsim N_0\, e^- \left(\frac{2.6}{0.026} \right)$$

$$N(\phi) \leqq 10^{-44}\, N_0$$

Clearly one would have to have an almost unimaginably large amount of metal at room temperature before there would be any reasonable chance of a single electron escaping due to thermal motion. Obviously this process is not practicable at room temperature. (We have rather badly misused the formula in obtaining this conclusion, i.e., ignoring the first term of the equation and the question of whether or not the electrons were traveling in the right direction, but the sheer magnitude of the result justifies our simplifications.)

However, if we raise the temperature an order of magnitude (ignoring the fact that most materials would have melted by 3000 K) we find the situation is more favorable because of the importance of the exponential term. At this temperature the number of electrons "hot" enough is roughly

$$N(\phi) \cong 4 \times 10^{-5}\, N_0$$

Since N_0 is of the order of Avogadro's number for a gram atomic weight of material, this number indicates that a ready supply should be available. This will remain so even after lowering the temperature to below the melting point of the material, and including the terms we ignored in the estimate. It is definitely possible to boil off measurable numbers of electrons from very hot metal surfaces. This process is called *thermionic emission*.

2.2 BASIC ELECTRON DEVICES For electrons to be of practical use, however, we must be able to get them away from the metal. If they are ejected directly into air, at a very short distance they will encounter a molecule whose mass is several thousand times their own and undergo a large-angle deflection, effectively stopping them (see sec. 2-6). If we wish them to travel any great distance, we must ensure that they travel in a vacuum such that their mean free path between collisions with gas molecules is much greater than the distance we want them to go. Once enclosed in a vacuum envelope it is quite straightforward to simply heat a piece of metal and collect the flow of electrons at some convenient point. The stream of electrons can be accelerated or decelerated by means of an electric field, and they may be deflected in any desired direction with a transverse electric or magnetic field, as indicated in Fig. 2-4.

2.3 ELECTROSTATIC LENSES A perhaps more interesting possibility for electron beams is the idea that they might be used in an analogous way to light rays in order to make optical devices. It is possible to make electron lenses, both with electric

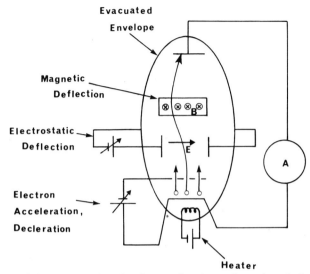

FIGURE 2-4 Vacuum tube, showing acceleration and steering of electrons.

fields and with magnetic fields. Since electrons are particles, not waves,* and electron "optical" device will not be affected by diffraction effects, so that it should be possible to design magnifying devices with much higher magnification than possible with the analogous optical instrument.

The analogy between electron beams and light rays can be seen very easily by considering the arrangement shown in Fig.2-5.

If an energetic electron traveling at a velocity v_1 passes through a region where there is a potential gradient, that is, an electric field, it will be accelerated in the direction of the gradient, but not in a direction perpendicular to it. This means that the following must be true

$$v_1 \sin \theta_1 = v_2 \sin\theta_2$$

where θ_1 and θ_2 are the initial and final angle of the electron trajectory relative to the potential gradient and v_2 is the final electron velocity. If made about light rays, that

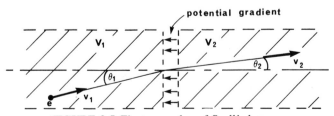

FIGURE 2-5 Electron analog of Snell's law.

* We will have to alter this notion later. See sec. 3-4.

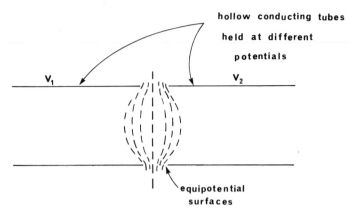

FIGURE 2-6 Electrostatic lens formed by shaped equipotential surfaces.

statement would have been called Snell's law; in this case it means that all the geometrical optics that have been worked out from Snell's law for light rays must also be applicable to electrons. For example we can regard equipotential surfaces as roughly analogous to an interface between optical media with different indices of refraction, that is, lenses.

In fact, on the strength of that analogy we would *guess* that an electrostatic lens could be made by placing two hollow conductors close together and setting up a potential difference $V_1 - V_2$ between them.

The equipotential surfaces between the two tubes will bulge as shown in Fig. 2-6, because a free charge near the tube wall will experience a force that is more nearly perpendicular to the lens axis than when near the center of the tube. This difference is due to the charge-free region at the end of the tubes. The length of the tube determines the magnitude of the influence of the "blank spots" at the end and hence the curvature of the equipotentials. Infinitely long tubes would produce parallel equipotential surfaces.

This picture certainly looks as though the arrangement should provide an electron lens, but to confirm this it is necessary to do at least some simple calculations. First

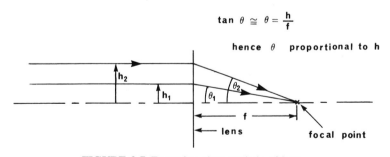

FIGURE 2-7 Focussing characteristic of lens.

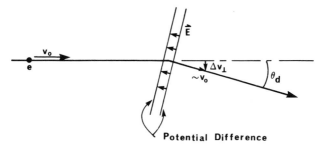

FIGURE 2-8 Electron deflection by potential difference.

we need to have a good working definition of an optical lens. It is a focusing device; that means for small deviations that the angular deflection of a ray passing through the lens is proportional to the distance from the axis of the lens as illustrated in Fig. 2-7.

The angle of deflection for electrons that pass through a region with a small potential difference can be worked out easily, and is shown in Fig. 2-8.

The deflection angle θ_d (assuming a small potential difference and a small deflection) is given by

$$\theta_d \cong \frac{\Delta v_\perp}{v_0}$$

where $\Delta v_\perp$ is the change in velocity perpendicular to the initial direction and v_0 is the initial velocity. This is related to the potential gradient and in terms of the component of the electric field at right angles to v_0, $E_\perp$ is given by

$$\frac{q}{m} E_\perp = \frac{\Delta v_\perp}{\Delta t}$$

so that

$$\Delta v_\perp = \frac{q E_\perp \Delta t}{m}$$

where q and m are the electron charge and mass, respectively. The time Δt required to pass through the potential difference, if it is a length l, is approximately

$$\Delta t = \frac{l}{v_0}$$

so that the deflection angle is

$$\theta_d = \frac{\Delta v_\perp}{v_0} = \frac{q E_\perp l}{m v_0^2}$$

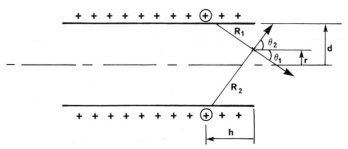

FIGURE 2-9 Electrostatic forces on a point charge in a charged hollow conductor.

Consequently, for this electrostatic device to act as a lens the perpendicular field component must be proportional to the displacement from the "lens" axis. Is such a condition actually true for the geometry we have chosen? To check whether or not it is true for a real three-dimensional system is unnecessarily complicated; a two-dimensional calculation should suffice to show the principle.

We can start with one point of the two-dimensional picture shown in Fig. 2-9 and calculate the electric field along the radial direction E_r due to a typical "pair" of charges on the two plates which have the same horizontal displacement h from the test point. It is

$$E_r = \frac{-kq \sin \theta_1}{R_1^2} + \frac{kq \sin \theta_2}{R_2^2}$$

$$= kq \left\{ \frac{d + r}{[h^2 + (d + r)^2]^{3/2}} - \frac{d - r}{[h^2 + (d - r)^2]^{3/2}} \right\}$$

where d is the distance from the axis to either plate and r is the distance from the axis to the point where E_r is calculated. For small displacements from the axis we can take $r \ll d$ so that

$$E_r \cong kq \, \frac{d + r - d + r}{(h^2 + d^2)^{3/2}}$$

$$\cong \frac{2k \, q \, r}{(h^2 + d^2)^{3/2}}$$

Therefore the radial field is proportional to the displacement from the axis and the device will in fact operate as a lens, as we guessed. (The same result would be obtained if we used negative charge on the cylinders instead of positive, the only change being that of the overall sign of the field.) To do a full calculation we would have to include all the charge on the plates (or in three dimensions on the cylinders), but clearly that will not affect the conclusion that E_r is proportional to r, which is the crucial requirement.

There is, however, an additional complication that must be examined, and that is the following: The lens arrangement is completely symmetric; while the electrons are forced toward the axis in the first half of the field region, the second half has a field configuration that looks as though it ought to deflect the electron right back to its original path, removing all focusing effects! But this is not quite true. Indeed the electron will be deflected in the opposite direction in the second half of the lens, but during the passage through the first half of the lens the electron has had a net acceleration. Its velocity has changed, and therefore when it enters the second half, even though the electric field distribution is the same as before, the effect on the electron is different since the deflection angle depends strongly on the velocity. Our electrostatic lens is in fact a converging-diverging lens doublet, but due to the change in electron velocity a net convergence or divergence is possible.

Having seen that this device can in principle operate as a lens, we should think of the limitations to its expected operational conditions. The obvious limitation that was built in from the start is the small deflection angle condition. This is certainly true, but can be improved upon by careful redesign of the lens geometry. The lens is also highly "chromatic." The focus point depends strongly on the electron velocity, so that monoenergetic electron beams must be used. This same energy behavior of the lens arrangement also puts a practical limit on the velocity for which the device will usefully act as a lens. In order for there to be a net focusing effect a significant change in electron velocity must occur while passing through the lens. Consequently the device becomes impractical for high velocity beams of electrons, and we must look to other effects to provide electron focusing.

2.4 MAGNETIC LENSES   From the relation $\mathbf{F} = q\mathbf{v} \times \mathbf{B}$ it is clear that the magnetic force takes on increased importance at high velocities, and we rightly would expect the magnetic field to become useful in high-velocity electron lenses. The main kinds of magnetic lenses are the quadrupole lens and the axial field magnetic lens. Of these the simpler type is the axial lens, and we will only consider its behavior. (For a description of the quadrupole lens see the references listed at the end of this chapter.)

To see whether or not a uniform axial magnetic field can possibly act as a lens we can consider the following. If we take a monoenergetic point source that is emitting a beam of electrons that are not diverging greatly (paraxial ray assumption) and immerse it in a uniform magnetic field directed along the mean beam direction, two things will happen. First, since $\mathbf{F} = q\mathbf{v} \times \mathbf{B}$, the component of the electron velocity parallel to B will be entirely unaffected. The component of velocity perpendicular to B will generate an acceleration perpendicular to itself and to B. This will give rise to uniform circular motion in a plane perpendicular to B which has an angular frequency

$$\omega = \frac{qB}{m}$$

The net result will be a spiral (Fig. 2-10) whose radius is determined by the component of the electron velocity perpendicular to B, but whose period is independent of that

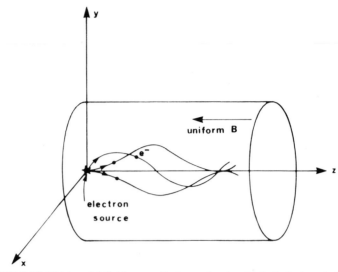

FIGURE 2-10 Thick axial field magnetic lens, showing typical electron trajectories.

velocity and depends only on q, B, and m. This is true for all rays emanating from the point source. They will all arrive back at the z axis once every cycle (which is the same value for all of them), and this is independent of their starting direction. Furthermore, since we have taken only electron trajectories with small deviations from the B direction and uniform velocities v_0, all of these electrons will cross the axis at a distance

$$S = vT$$

which is the same for all of them, since v is approximately the same and equal to v_0, and T, the time to complete one cycle, is the same. All the electrons will converge at a point a distance

$$S = \frac{v_0 2\pi m}{qB}$$

from the source position. They have been *focused*. This uniform axial field clearly is acting as a thick magnetic lens. (The lens is *thick* because the electrons have their entire trajectory inside the lens.)

It is also possible to make a thin lens with an axial magnetic field. A complete calculation with a realistic form of axial magnetic field necessary for an axial field lens is rather more complicated than is useful for our purposes. However we can convince ourselves that a uniform axial magnetic field satisfies the requirement laid down previously for a device to act as a lens. To start with we will make the mathematically convenient assumption of a thin disc of uniform magnetic field of strength B and thickness l (Fig. 2-11), and a source of monoenergetic electrons a distance x

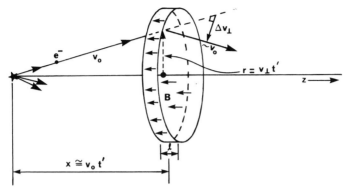

FIGURE 2-11 Ideal thin axial field magnetic lens, showing electron trajectory and associated geometry; t' is the time required for the electrons to reach the lens.

away. (Only electron trajectories close to the z direction will be considered.) The electron velocity v_0 entering the lens can be broken down into a component parallel to the z axis, $v_\parallel$ (which is assumed approximately the same for all electrons), and a component perpendicular to the z axis, $v_\perp$. Due to $v_\perp$ the electrons will execute uniform circular motion in the plane perpendicular to B when they are in the magnetic field region. That is, the magnitude of $v_\perp$ will remain constant, but rotate through some angle θ given by

$$\theta = \omega t$$

where

$$\omega = \frac{qB}{m}$$

and t is the time spent in the uniform magnetic field. This rotation causes a change of the perpendicular component of the velocity

$$\Delta v_\perp = v_\perp(1 - \cos \omega t)$$

which has the effect of a net deflection of the electron back toward the z axis plus a rotation about the axis of the lens after passing through the magnetic field. From this we can obtain a deflection angle analogous to the electrostatic case

$$\theta_d = \frac{\Delta v_\perp}{v_0}$$

where now

$$\Delta v_\perp = v_\perp (1 - \cos \omega t)$$

For small ωt (thin lens assumption)

$$\cos \omega t = \sqrt{1 - \sin^2 \omega t}$$

$$\cong 1 - \frac{\omega^2 t^2}{2}$$

so that

$$\Delta v_\perp = v_\perp \left(1 - 1 + \frac{\omega^2 t^2}{2}\right)$$

$$= v_\perp \frac{\omega^2 t^2}{2}$$

The time t taken to pass through the lens region is given by

$$t = \frac{l}{v_0}$$

and the deflection angle is therefore

$$\theta_d = \frac{v_\perp \omega^2 t^2}{2v_0}$$

$$= \frac{v_\perp}{v_0} \frac{q^2 B^2 l^2}{2 \, m^2 v_0^2}$$

The deflection angle has to be related to the displacement from the z axis. This is given by the relation

$$r = v_\perp t'$$

where t' is the time it takes the electron to reach the lens from the source and is related to the displacement of the lens from the source x, by

$$x \cong v_0 t'$$

so that

$$\theta_d = \frac{r}{x} \frac{q^2 B^2 l^2}{2 m^2 \, v_0^2}$$

Therefore the deflection is proportional to the displacement from the axis, and the magnetic field arrangement will in fact operate as a lens.

In attempting to apply this formula to a real axial magnetic field lens, we would discover that there are very important corrections to the lens behavior due to the fact that there are always nonaxial fringing fields present. In a precise, quantitative calculation of the properties of axial magnetic lenses, as indeed for electrostatic lenses, it is most likely that a computer-calculated tracing of the electron trajectories would be carried out.

2.5 ELECTRON MICROSCOPES   With the aid of a source of free electrons and the various types of lenses to converge or diverge beams of electrons, we are in a position to make electron analogs of optical devices. The optical device that is of most interest to us is the microscope. This device is limited in its magnification by diffraction effects of the electromagnetic waves that comprise light. Such a limit should not exist for electrons, as they are particles and indeed simply travel in straight lines (in the absence of a force) not subject to diffraction effects.

Let us remind ourselves how an optical compound microscope works. This is shown in Fig. 2-12 together with an equivalent, but somewhat more cumbersome, configuration.

The latter arrangement has the advantage that the final image is real. Also illustrated is a condenser lens, or device to throw as much light as possible onto the object. The object blocks light rays in a pattern corresponding to its shape, and the pattern of

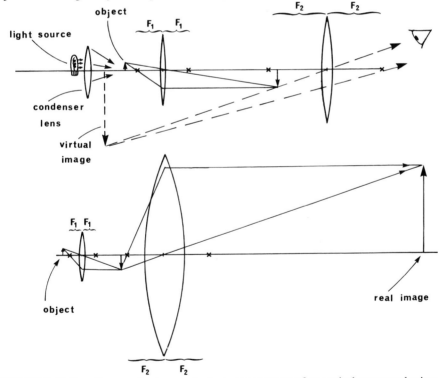

FIGURE 2-12 Ray diagrams for two equivalent arrangements of an optical compound microscope.

emerging rays is magnified by the lens. The greatest magnification we can expect from such a device is roughly the ratio of what we can resolve with the unaided eye, ~ ¼ mm, to the diffraction limit of light, ~ ¼ μm, a factor of 1000 or so.

The analogous "optical" device using an electron beam, illustrated in Fig. 2-13, does not suffer from this limitation. Electron microscopes are able to magnify objects from 30 to 100,000 × (they can resolve structure down to $\approx 3 \times 10^{-4}$ μm). The limitation on magnification appears to be aberration problems with the lens system and mechanical vibration effects. There are other aspects of electron microscopes that provide limitations of a different nature. First, of course, the object being magnified must be able to exist in a vacuum. This is not an overwhelming problem, but does restrict somewhat the objects that can be viewed.

The second problem comes from the fact that the electrons must pass through the object to be magnified (hence the name transmission electron microscope, TEM) and the relative numbers able to pass through different regions of the object provide the contrast in "opacity." This means that the object, or sample, must be very thin (less than about 0.1 μm thick, typically) and hence requires special preparation. It also means that the electrons must be energetic. The electrons are accelerated to kinetic energies ranging from several tens of keV to a few MeV, depending on the material to be viewed. This in turn gives rise to another problem: object heating. As shown in Fig. 2-14, the "opaque" regions of the sample can have all the electron kinetic energy deposited in them and can therefore become so hot that they evaporate. This depends on beam intensity and energy and target thickness, but frequently provides a

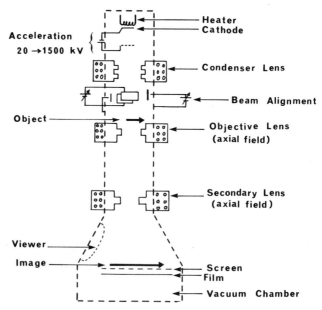

FIGURE 2-13 Schematic diagram of a transmission electron microscope.

practical limit to transmission electron microscopy due to object damage or destruction.

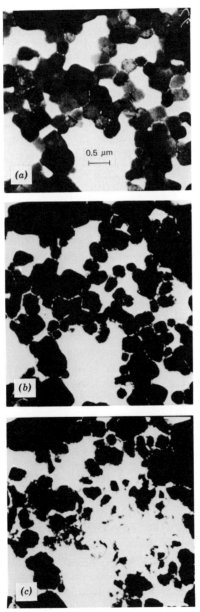

FIGURE 2-14 A thin sample of potassium iodide crystals being examined with a transmission electron microscope; a low intensity exposure (*a*), and high intensity sequence (*b*) and (*c*), showing target deterioration due to electron heating of the sample.

The limitations of the TEM—excessive power dissipation in the sample and the necessity of a semitransparent material—led to the development of an electron device similar in purpose but considerably more simple in design and versatile in use. This is the scanning electron microscope, or SEM. With this device, shown in Fig. 2-15, the only "optical" requirement is a finely focused beam of electrons that strikes a target placed at its focal point. The beam is then scanned across the target surface, normally in a X-Y (raster) scanning mode, by a controlled X- and Y-deflection magnet. The control for the X and Y deflection is also used to sweep synchronously the electron beam of a cathode ray tube (CRT). The intensity on the CRT can then be controlled by any one of a number of types of signals from the target. The magnification of the microscope is determined by the ratio of the scan size of the electron beam in the SEM to that on the CRT. That this device can be used to "see" the contours of the sample (which no longer has to be thin) can be realized readily by considering the intensity of scattered electrons in a detector if the beam goes on a flat surface, in a "hole," or behind a hump.

If the output of the electron detector is connected to the intensity of the CRT display, the output will be a "picture" of the physical condition of the target, magnified by powers slightly less (resolution $\geqslant 0.02\ \mu$m) than that available for the TEM. The SEM, however, has much greater versatility, as the intensity control can be obtiined not just from back-scattered electrons, but from secondary emission electrons, light emission, X-ray emission, or simply the current to ground through the target. All of these signals are determined by different aspects of the material being investigated, and therefore give information about atomic, chemical and electrical properties of the materials in addition to their physical properties. An example of the use of an X-ray detector is shown in Fig. 2-16 where the same section of material has been scanned simultaneously for concentration of four different elements.

With the SEM most of the problems of target preparation and heat dissipation have been overcome. As the electrons do not have to pass through the material, the target does not have to be thin, and the electron energy does not need to be so high. This considerably reduces the heat dissipation problem.

2.6 SECONDARY ELECTRON EMISSION   One of the phenomena observed with the SEM is *secondary electron emission* when the main beam of electrons strikes the metal target surface. To understand secondary electron emission and why it might occur we must ask ourselves what happens when an energetic electron "strikes" the surface of a metal. The electron will "collide" with (exert a force on and therefore share energy with) either the free electrons in the metal or with the atoms making up the lattice of the metal. If they strike the latter they are likely only to change directiion and, because of their small relative mass, lose practically no energy. However when they strike an electron they will lose anywhere from zero energy to all their energy,

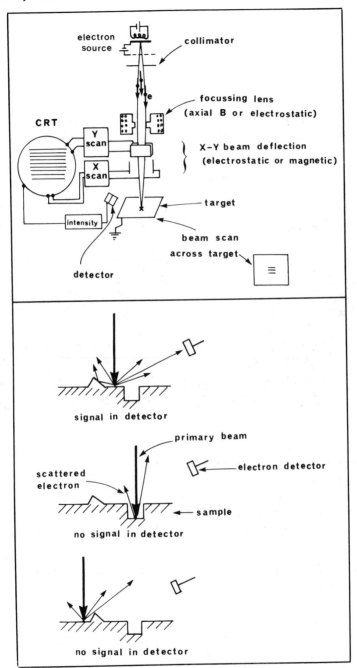

FIGURE 2-15 Schematic diagram of scanning electron microscope, also showing method of detecting surface contours.

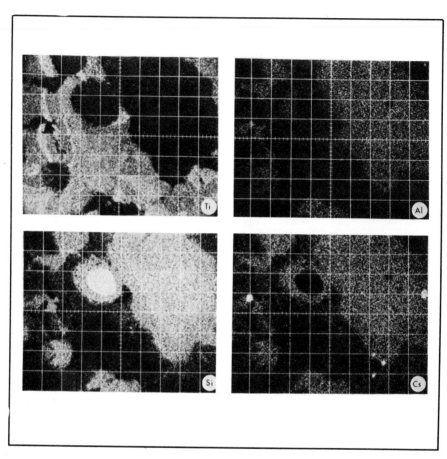

FIGURE 2-16 Example of scanning electron microscope used with X-ray detector. Four separate elements are shown in identical scans of a titanate ceramic waste form taken to identify the form of cesium binding. The grid size is 50 microns. From R.S. Claassen, *Physics Today* 29, number 11, p. 23 (1976). Reprinted by permission, Sandia Laboratories, Albuquerque, New Mexico.

depending on the details of the collision*. On the average we might expect them to lose approximately half their energy.

If we were to start with an electron of moderate energy (say 100 eV) striking the surface of the target, very quickly we would have a "shower" of energetic electrons, as illustrated in Fig. 2-17; *some* of them will be capable of escaping from the surface (assume $\phi \simeq 4$ eV). The question now becomes how many? Let's estimate a *most favorable case* for secondary emission. Assume that no collision gives an electron of less energy than ϕ; then eventually we would have

* See Appendix 2

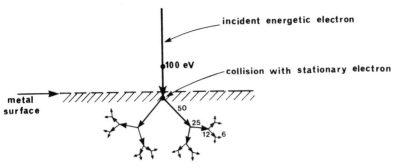

FIGURE 2-17 Production of secondary electrons in a metal.

$$\frac{eV_0}{\phi} = \frac{100}{4} \cong 25 \text{ electrons}$$

with enough kinetic energy to escape. What is the probability that these energetic electrons can escape? Optimally, it is the probability that they are going in the right direction. If we assume that these electrons are at the metal surface, approximately half of all directions are "right directions." This suggests that there should be an upper limit of

$$\approx \frac{25}{2} = 12 \text{ electrons}$$

that could be emitted from the surface after the single energetic electron enters the metal. This is a very optimistic calculation; is it anywhere near the truth? In fact we would find that something like three to four secondary electrons are emitted from typical surfaces. We should not be surprised at our overestimate for several reasons:

1. Energy can be lost through electrons with $KE < \phi$.
2. Electrons far in from the surface are unlikely to reach the surface with $KE > \phi$.
3. Electrons can come out with excess KE, reducing available energy.
4. Possibility of other energy-loss mechanisms reducing available energy.

Experimentally the number of secondary electrons emitted depends on the primary energy and the work function of the material. A maximum in secondary emission is found for $\sim$ 400 to 500 eV electrons, and the number emitted per primary can be anywhere from something like 1.2 (for Ni) to 8 (for Cs with a special surface coating). What is of major significance to us, however, is that we have found a process that is capable of multiplying—amplifying—electron currents.

2.7 ELECTRON- AND PHOTO-MULTIPLIERS The potentially useful fact about secondary emission is that it is possible to increase the number of electrons in a beam by means of this effect. The increase possible at first sight appears small, but when it is recognized that the secondary electrons themselves may be used to cause additional secondary emission, it is clear that the amplification possibilities are virtually limitless.

In order to achieve large electron multiplication it is only necessary to ensure that the secondary electrons from an initial primary electron are accelerated through a large enough potential difference so that they in turn will cause a further stage of secondary emission. This process can be repeated many times.

As illustrated in Fig. 2-18, each time it happens *each* of the arriving electrons will cause an increase by the same factor (assuming the potential difference between stages is the same) so that the total multiplication factor M (if the number of secondaries per primary is δ) for *n* stages of multiplication is

$$M = \delta^n$$

This means that if a single electron strikes the first stage of a 10 stage, δ = 4 device, 10^6 electrons will leave the last stage! We clearly have a simple, yet effective, high-gain charge amplifier. Such devices do exist with gains of 10^6, although in practice it is necessary to ensure that both a high number of secondaries per primary are emitted (low-ϕ materials) and that, through either electrostatic or magnetic focusing, few of the electrons are lost in the multiplication chain.

Let's look at some of the other characteristics we might expect of an electron multiplier. For instance, if a single electron strikes the first electrode, how long does it

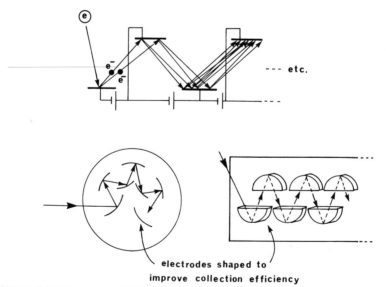

electrodes shaped to
improve collection efficiency

FIGURE 2-18 Electron multipliers. Electrode geometry can be arranged for maximum secondary emission electron collection efficiency.

take for the amplified charge pulse to form? For ease of calculation we can assume a constant electric field between all electrodes, a separation s between the plates of 1 cm and a potential difference of 100 volts between adjacent plates. Then the transit time t between adjacent electrodes is given roughly by

$$s = \frac{at^2}{2}$$

so that

$$t = \sqrt{\frac{2s}{a}} = \sqrt{\frac{2s\,m}{qE}}$$

$$= \sqrt{\frac{2 \times 1 \times 10^{-2} \times 10^{-30}}{1.6 \times 10^{-19} \times 10^4}}$$

$$= 3 \times 10^{-9}\,s = 3\ ns$$

The total transit time for a 10-stage electron multiplier should be about *30 ns*. The current would increase of course, as the charges passed from stage to stage. From the fact that the increase in number N of electrons as they pass through each stage is given roughly by

$$\frac{\Delta N}{\Delta t} \cong (\delta - 1)N$$

we can expect an exponentially increasing current pulse, as shown in Fig. 2-19. We can also estimate the maximum current, which will be where the current is passing from the next-to-last stage to the last stage. A rough guess would be that the current at that stage is the charge arriving at that stage divided by the transit time.

$$i \cong \frac{Q_f}{\Delta t} = \frac{10^6 \times 1.6 \times 10^{-19}}{3 \times 10^{-9}} = 50\ \mu A$$

Depending on the actual field variation a proper calculation would confirm such a simple estimate. Even then the results would be only approximate; a precise calculation

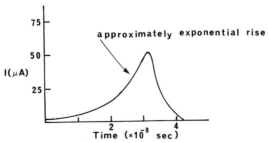

FIGURE 2-19 Approximate form of current pulse for output of electron multiplier.

would require a detailed knowledge of the transit time variation which in turn depends on the detector geometry.

This amplification technique is most frequently used in a device called a *photomultiplier*. It is possible to start the initial stage of the device using light instead of moving electrons as a source of energy. Indeed we easily can think of this as just a way of pumping energy into a piece of metal using electromagnetic radiation in the form of light.

Light is a transverse oscillating electric and magnetic field, and it can give kinetic energy to electrons and allow them to escape from the metal. The whole process is reminiscent of the thermionic emission which we discussed previously. This parallel can be recognized easily and be put to use, without even asking whether such an assumption is warranted or fundamentally correct (which we certainly must do ultimately).

It has been known experimentally for some time that if the arrangement for a normal vacuum-tube diode is modified by coating the cathode surface with a low work-function material such as the alkali metal sodium, electrons will flow from the cathode to the anode when the cathode is illuminated by visible light. As only electrons are ejected from the cathode, current can only flow in one direction. Such a device is called a *photodiode* (Fig. 2-20) due to the close analogy with thermionic vacuum tube diodes. By incorporating photodiode into the first stage of an electron multiplier we can convert it into a *photomultiplier,* which is sensitive to levels of illumination several orders of magnitude lower than can be detected with photodiodes.

The electron multipliers discussed so far have no position-sensing capability. A recently developed device, called a microchannel plate, is capable of providing position information with a resolution of the order of 50 μm, as well as electron multiplication comparable to the normal electron multiplier. The microchannel plate, as shown in Fig. 2-21, consists of a set of closely spaced tubes, or microchannels, which have a diameter ranging from 15 to 50 μm and are distributed throughout a semiconducting

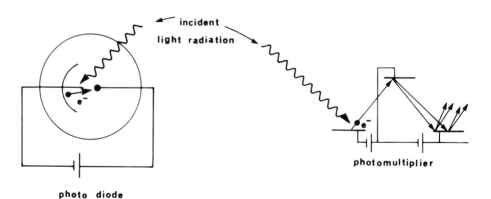

FIGURE 2-20 Photodiode and photomultiplier.

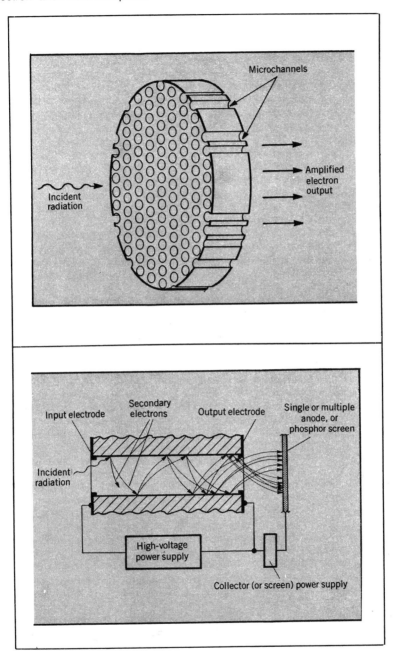

FIGURE 2-21 Arrangement and operation of microchannel plate. Details of operation are discussed in the text. From B. Leskovar of Lawrence Berkeley Laboratory in *Physics Today* 30, number 11, p. 43 (1977).

glass medium. Channel separations range from 20 to 60 μm. A potential difference of approximately 1000 volts is maintained across the length of the tubes and incident radiation starts a cascade of secondary electrons traveling down the length of the channel, which is 0.5 to 2 mm long. The initial electron is generated either by striking the wall of the channel directly or by means of a photo-electron converter placed in front of the microchannels. The arrangement is capable of providing a gain of 10^3 to 10^4. The output electrons from a channel can be allowed either to strike a phosphor screen directly, producing a visual image of the source, or can be passed through additional sets of microchannel plates to produce overall gains of up to 10^7, and then handled by standard signal processing techniques.

Because of the small channel length and high applied voltage, the total electron transit time is much shorter than for a conventional electron multiplier with the same gain. Pulse risetimes are characteristically less than 10^{-9} s per plate with a spread in rise-times of the order 10^{-10}s, so that the microchannel plate can also be used as an extremely high resolution timing device.

FURTHER READING   N. W. Ashcroft and N. D. Merwin, *Solid State Physics,* Chapter 18 (Work Functions), Holt, Rinehart and Winston, 1966.

J. Birks, *The Theory and Practice of Scintillation Counting,* Pergamon, 1964.

J. M. Cowley and S. Iijima, Electron Microscopy of Atoms in Crystals, *Physics Today,* March 1977.

P. S. Farago, *Free-Electron Physics,* Penguin, 1970.

E. W. Müller and T. T. Tsong, *Field Ion Microscopy Principles and Applications,* Elsevier, 1969.

C. W. Oatley, *The Scanning Electron Microscope,* Cambridge University Press, 1972.

PROBLEMS   1. An electron field of 1×10^9V/m is required for field emission of electrons. If a field emission device is made from two concentric spheres with a potential difference of 4000 volts between them, and the outer sphere has a radius of 3 cm, what must the radius of the inner sphere be in order to achieve field emission?

2. From the description that we have used for a free electron inside a metal how do you expect the work function to be related to the metal's temperature? What is the observed relation between temperature and work function, and what conclusion can you draw from that fact?

3. Calculate the fraction of a free electron gas at 800 K which has enough kinetic energy to escape from a metal with $\phi = 2.5$ eV using the full Maxwell-Boltzman distribution.

4. Prove that the maximum of the Maxwell-Boltzman energy distribution occurs at $kT/2$. What is the average energy (in terms of kT) of this distribution?

5. Discuss the desirable temperature and work function characteristics of a filament for an electron microscope. If you had to design an "improved" filament what characteristic would you work on first?

6. Estimate what the current density of electrons leaving the surface of a metal that is heated to a temperature of 400°C should be. Assume that $N(\phi) = 10^{-6}N_0$ and that the density of free electrons in the metal is $10^{23}/cm^3$. Treat the electrons as if they were all traveling at the same velocity, given roughly by $E = kT$, that the probability of traveling in the right direction to escape is 50 percent, and that the probability per electron of escape is given by $N(\phi)/N_0$. Does your estimate correspond to what is observed under these conditions? How reasonable are the various approximations that have been made?

7. (a) An electrostatic focusing device is arranged so that it has an electric field perpendicular to the beam axis of the following form

$$E(r) = 2 \times 10^1 \, r - 2.5 \times 10^{-2} \, r^3 + 7.8 \times 10^{-3} \, r^6$$

in the radial direction, where r is the distance from the beam axis in cm. What is the maximum useful diameter of the lens (*i.e.*, less than 10% error in the beam deflection angle) for a distant electron source?

   (b) If a magnetic lens put at the same point is just capable of providing the same focal length for 30 keV electrons as the above electrostatic lens, which of the two devices should be able to provide a shorter focal length (smaller image distance) for 10 keV electrons, and why?

8. A uniform monoenergetic electron beam for an electron microscope is to be collimated by a large sheet of high-resistance material with a circular hole at the center. During operation it is found that the sheet becomes charged to an approximately uniform potential. You are asked to determine theoretically whether or not this will act as a focusing device. Give the conditions necessary for the hole to act like a lens (do not *solve* the problem—it's too difficult). Do you think it will act like a lens? Why?

9. Many problems can be solved by the simplest possible consideration. From the known basic behavior of electrons in a uniform magnetic field and the symmetry of the arrangement, determine the minimum possible value of B for a 1 cm thick, ideal axial magnetic lens using 10 keV electrons if object and image are both 50 cm from the lens.

10. You have been asked to choose a vacuum system for a new SEM that is being designed. You know that on the average atomic radii are about 2 Å and you can estimate the atom density in a gas at a pressure P from the ideal gas law. From these two facts work out what the probability per unit length is for an electron to collide with a gas atom, and from that derive a value for the *average* distance electrons travel before colliding with an atom (mean free path) as a function of pressure. If the electrons have to travel 1 m in the SEM, what pressure would you need to ensure that the mean free path is greater than 1 m?

11. An electron gun has an accelerating voltage of 400 V and an axial magnetic focusing field. If the target is 10 cm from the source, what must the minimum value magnetic field be to focus the electron beam on the target? What other values should also work?

12. Electrons are to be focused on a target using a thick, uniform B axial magnetic lens. If the source-target distance is 30 cm and the maximum field strength available is 0.04 Tesla, what is the maximum energy electrons can have if they are to be focused in this device?

13. From what you know about the process of secondary emission, compare the sensitivity and spatial resolution to be expected for an SEM when direct backscattered electrons are used to form an image to that expected when secondary emission electrons are used.

14. What "spot size" do you expect when using secondary emission for SEM imaging? Assume one available electron per atom, and an incident 1.5 keV beam which produces secondary electrons with an average energy approximately 100 eV. Estimate a free electron collision "size" (see Appendix 1) of πb^2, where b is the impact parameter for collisions that result in scattering through at least 15° (see Appendix 2). Use the approximations in the text for secondary electron production and the results of problem ten to arrive at your answer. Do you expect your answer to be an underestimate or overestimate? Why? How does it compare with what is observed?

15. What is the uncertainty in the total charge out of a 10-stage multiplier with a single stage gain of 5 for a single electron input? Assume that the multiplication is a completely statistical process, that is, that the fluctuation in the secondary electron yield is the square root of the *average* yield, and that there are no correlations between the stages. If it were possible to change the gain of one of the stages, which one would be the best to improve the accuracy, and why?

16. Using the basic model of secondary emission that was developed, how do you expect the secondary electron yield to vary with incident electron energy? What is the cause in the reduction of yield at very high energies?

17. You have been asked to design the multiplication stages of a standard electron multiplier so that the stage gain, already optimized for primary electron energy and collection geometry, can be increased significantly. Assuming that it can be arranged so that any emitted secondary can be collected, can you think of a modification to the plate that, in principle, could do this?

18. Estimate the transit time for electrons traveling down a 2 mm long straight, narrow tube which is in a uniform electric field of 10,000 V/cm. Assume that five uniformly spaced collisions with the walls occur.

19. Roughly estimate what the transit time spread would be in problem 18 if there were five collisions but only uniformly spaced *on the average*.

THREE

BEGINNINGS OF MODERN PHYSICS

3.1 THE PHOTOELECTRIC EFFECT If we recognize the fact that by either heating a metal or irradiating it with light we basically are giving electrons excess kinetic energy, then we immediately would expect that all the properties of thermionic emission should also occur for photoemission. Certainly we would expect that any material that emitted electrons in the thermionic effect should also emit electrons under irradiation by visible light.

However, this is not so. Only certain materials, such as sodium and potassium which have work functions on the order of 1 eV, are found to exhibit photoemission. In fact there are several characteristics of photoemission that are markedly different from thermionic emission and the careful study of these properties will serve to lead us to new concepts of physics.

Several other characteristics of the photoelectric effect, shown in Fig. 3-1, require further attention and detailed comparison with the behavior in thermionic emission. The time development of the current flow in photoemission characteristically reaches a steady value in no more than about 10^{-9} s after the start of irradiation, regardless of the light intensity, even for intensities as low as 10^{-10} W/m². This is to be compared with the onset time required for thermionic currents, which with heating of the order of 3 watts over $\sim$ 3 cm² ($\sim 10^4$ W/m²), have a characteristic rise time of $\sim$ 30 s.

For a given frequency of incident light there is a well-defined maximum kinetic energy of the ejected electrons *which is independent of the light intensity*. For thermionic emission the higher the power input, the higher the temperature is, and therefore the higher the maximum electron kinetic energy.

For photoemission the maximum electron kinetic energy* is found to be related linearly to the frequency v of the incident light. This linear relation is the same for all

---

* We can assume that these electrons were emitted directly from the surface and lost no energy through collisions inside the metal.

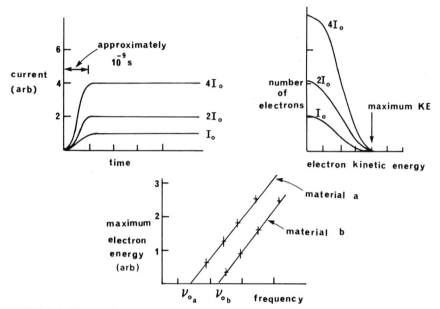

FIGURE 3-1 Characteristics of photoelectric emission for different light intensities, and emitting materials.

materials but the frequency at which KE_{max} goes to zero (photo current ceases) is different for different materials. Furthermore, below the cutoff frequency no current is observed regardless of the light intensity.

For a thermionic process, there should only be a relation between total power input and electron energy. The electron kinetic energy should not be related to radiation frequency, so there should not be a cutoff frequency.

Clearly almost every experimental aspect of the photoelectric effect—the emission of electrons as a result of irradiation by light—is in complete disagreement with what we would have expected to happen. The rate of energy delivery to the photocathode is unrelated to the time required for current to start flowing; the maximum electron kinetic energy is unrelated to the radiation intensity (which it should be) and is directly related to the frequency of the incident radiation (which it should not be). When the situation becomes as apparently contradictory as this, the best thing to do is to drop all previous ideas about what is happening and rely only on the experimental facts.

All we really know is that when light shines on the photocathode, energy is transferred to the electrons in the material in some way, the details of which are not known to us. We know more than just that, however. We know that the energy source

    1. causes electron ejection "instantaneously";
    2. causes ejection for any energy flux; and

3. the maximum electron kinetic energy is related only to the frequency of the light.

What ways of delivering energy are consistent with this behavior? There are really two ways of delivering energy; *via* waves or particles. Let's compare how these two sources would deliver energy to a material (say four electrons bound into four atoms). We can arrange things to happen in a most telling way. Let the total amount of energy be delivered in a very short time, and be just enough so that if it is concentrated in one atom, the electron will become unbound (ejected). Then the picture according to the concept of wave irradiation would be that shown in the upper part of Fig. 3-2.

Here, since the energy is distributed continuously over the wavefront, the incident energy is delivered uniformly to all four of the electrons, so that no single one of them can escape, but they all have an increased kinetic energy. Only if the four electrons have *exactly* the right sequence of collisions would it be possible to concentrate enough

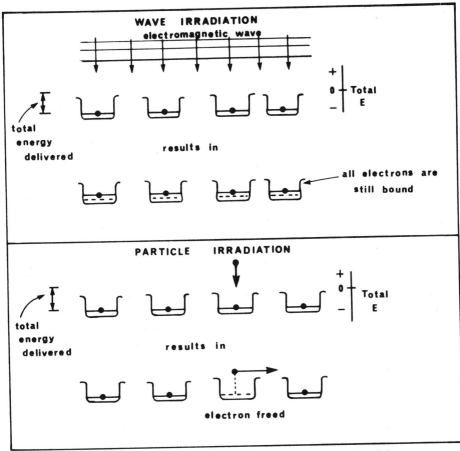

FIGURE 3-2 Energy delivered to a material via waves and particles.

kinetic energy in one electron for it to escape. This is rather unlikely for the four sample atoms we have chosen, and it will become increasingly more unlikely as the number of atoms approaches that appropriate to a real piece of metal.

The particle irradiation picture, as can be seen in the lower part of Fig. 3-2, is considerably different. Since the energy carried by particles comes in "lumps" it can be transferred immediately to a single electron, which is able to escape from the material. Furthermore this property is retained when the number of atoms is increased. It is also completely consistent with all the essential properties of photoemission as we have listed them. The photoelectric effect experiment clearly demonstrates that light is behaving like a stream of particles. The only consistent picture we can make is that when we shine light on an object we bombard it with a "rain" of energetic particles!

This conclusion is somewhat disturbing, mainly because of the known diffraction effects of light, which are specifically wave properties. The only comment that can be made is that the physical situation was different when it was concluded that light is a wave. There, wavelike behavior was observed *when light was forced to pass through narrow slits,* that is, it was localized. In the photoelectric effect no constriction was placed on the light path and *in that case* light acts as a particle. This makes nature appear more complicated than it has before, a situation reminiscent of special relativity. Just as we did not have the proper "feeling" for space or time, our intuitive feeling for light as a particle or a wave turns out to be untrustworthy. Its nature takes on either aspect in different experimental circumstances.

Special relativity itself, looked at properly, might lead us to wonder whether light ought not to act in some way like a particle. Since light is a form of energy and since we presume *all* forms of energy are interconvertible, we must expect that $E = mc^2$ can be applied to light as well as to physical objects. Therefore the energy content of light must be associated in some way with mass, even though it is likely not the way we are used to associating with particles. (We will return to this notion in sec. 3-4.)

If we are able to bring ourselves to accept a particle nature for light (the particles are called *photons*), several of the experimental results discussed above are explained readily. The instantaneous emission is a natural consequence, as is the presence of a current at all flux levels. The particle or photon energy is shown to us directly from the light frequency vs the electron kinetic energy curve. From the experimental results we see that the photon energy is related linearly to the frequency of the light. The slope, which we saw was the same for any photoemitting material, gives us the quantitative relation

$$E_{\text{photon}} = h\nu$$

where experimentally $h = 6.6 \times 10^{-34}$ Js, and is called Planck's constant. From this relation, which is entirely unexpected classically, we can see that there *should* be a threshold frequency, below which not enough energy is given to an electron to enable it to escape. This energy corresponds to the material's work function, that is,

$$h\nu_{\text{thresh}} = \phi$$

In fact if we take ν corresponding to a convenient visible light wavelength, $\lambda = 0.7$ μm (700 nm), we find that this corresponds to approximately 2 eV. This immediately explains why only low work-function materials exhibit photoemission. The range of visible light goes approximately from $\lambda = 0.7$ μm for red light to 0.4 μm (400 nm), for violet light. These correspond to photon energies of 1.7 eV for red and 3.1 eV for violet light.

All the experimental behavior of the photoelectric effect can be understood readily in terms of light being a flux of photons that carry an energy $E = h\nu$. We are then left with the question of why light should be this way. That is tantamount to asking what the mechanism of generating light is; to obtain any further understanding we must tackle that problem.

Notwithstanding our present ignorance at the level of the basic physics, there is a wide range of practical applications for this effect. Perhaps the most interesting is the use of the photoelectric effect to enhance thermionic electron emission of heated samples, which can then serve as an object in an electron microscope. Devices using this principle have been developed (see Fig. 3-3) and are called photoemission electron microscopes. The material to be viewed is irradiated with ultraviolet light and the image is formed by focusing the electrons emitted by the sample. When the incident light frequency only just exceeds the photoelectric threshold, the electron current is very sensitive to the physical and crystallographic properties at the sample surface.

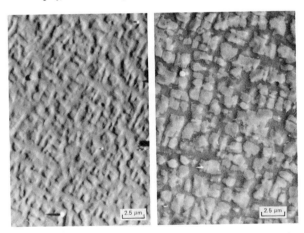

FIGURE 3-3 Photo-emission electron microscope image (*right*) and secondary emission image (*left*) of the same sample of high temperature alloy nimonic, taken at room temperature. Different metallurgical phases of the alloy can be discerned in the photoemission image, while the secondary emission image only shows topography. The secondary emission image was produced by low-angle bombardment with neutral particles. From L. Wegman, Photoemission Electron Microscopy, *High Vacuum Report* 23, September 1969, Balzer Ltd., Fuerstentum Liechtenstein.

If we are to even attempt to reconcile the particle and wave picture of light, we must look in some detail at the mechanism of generating light. The simplest classical way to generate a transverse electromagnetic wave (light) is by means of a modulated dipole charge, as illustrated in Fig. 3-4.

The size of the antenna is related to the wavelength of the radiation and should not be much greater than about half a wavelength to have a reasonable efficiency. For light, whose wavelength is on the order of 0.5 μm or less, the "antenna" should be no larger than this size. The size of an atom is on the order of 10^{-4} μm, so light radiation might be expected to arise from coherent motion of charges in systems anywhere from single atoms to several thousands of atoms. Before we attempt to understand coherent motion of thousands of atoms, clearly we must know what the motions of the components of *one* atom are. This we will look at next.

3.2 MODELS OF THE ATOM   The atom, as we know, is a small entity that consists of equal amounts of positive and negative charge (protons and electrons). What we must determine, however, is how these charges combine to form a stable atom. At the time this was an unanswered and serious question (about the turn of the century) there were two models of how protons and electrons might form atoms.

The simplest model (which we will call the Rutherford atom) was a picture analogous to the planetary system. For the simplest atom (hydrogen) an electron would circulate around the much more massive proton, attracted to it by the Coulomb force. In this picture it is tacitly assumed that the intrinsic "size" of the electron or proton would be much smaller than the orbit. For our purposes they can be considered point charges. From the equations of motion for the electron (mass m and charge e circling at a speed v and radius r) it is easy to see how this could lead to a stable atom. The total energy is given by:

$$TE = KE + PE$$

$$= \frac{mv^2}{2} + \frac{kq_p q_e}{r}$$

$$= \frac{mv^2}{2} - \frac{k e^2}{r}$$

The electron velocity and orbit radius are related through

$$\frac{mv^2}{r} = \frac{k e^2}{r^2}$$

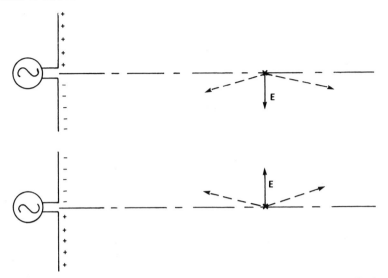

FIGURE 3-4 Generation of transverse electromagnetic radiation by oscillating dipole charge distribution; the alternating charge pattern produces an alternating transverse field.

which states that the centrifugal acceleration balances the attractive force. Hence

$$mv^2 = \frac{k\,e^2}{r}$$

and therefore

$$TE = \frac{k\,e^2}{2r} - \frac{k\,e^2}{r}$$

$$TE = -\frac{k\,e^2}{2r}$$

For a constant total energy (Fig. 3-5) the electron will continue to rotate at a constant speed and at a constant radius around the proton. The atom should be stable provided that total energy is conserved.

But is the total energy conserved in this system? No! A rotating electron-proton pair will look like a rotating dipole to a distant charge. It is effectively a dipole antenna. Such a dipole arrangement is certain to set up a transverse electromagnetic field and radiate away energy. When this happens, as it must, the total energy will decrease. The radius of the atom will then decrease. Furthermore, the electron will continue to rotate around the proton, so the radiation will continue. Consequently the radius of the

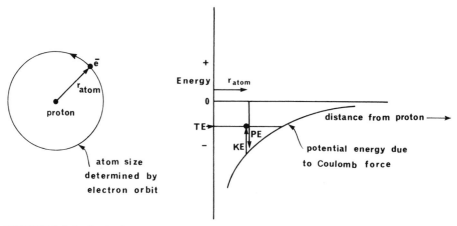

FIGURE 3-5 Rutherford model of the atom. Total energy, *TE,* is the sum of kinetic energy of the electron, *KE,* and potential energy, *PE*, due to attraction of the proton.

atom will decrease continuously as the atom continuously radiates off energy. The system will suffer a radiation collapse! Furthermore we would find that the collapse should happen in a time less than $\sim 10^{-8}$ s, if the relevant calculation was done. In view of the predicted radiation collapse for this model, it was felt to be untenable.

In order to avoid the collapse of the atom a *static* model was proposed, since it is the motion of the electron which caused the trouble. In this model, called the Thomson or plum pudding model (see Fig. 3-6), it was assumed that the *proton* gives the atom its size, and that the proton's charge is distributed uniformly throughout the volume of the atom. The electron, as before, is attracted to the proton and "collapses" into it. Inside the protonic matter the electron will come to rest at the center. If there are several electrons, they they will come to rest at equilibrium separations, balancing their mutual repulsion against the net central attraction of the "exposed" positive charge.

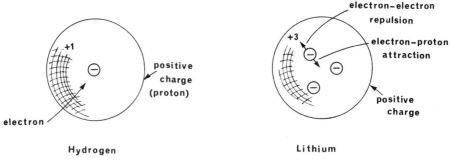

FIGURE 3-6 The Thomson model of the atom. In stable atoms all the charges are static.

This provides a consistent picture of the atom; there is no motion, hence no radiation. Large numbers of electrons can be accommodated inside the atom, and on somewhat closer examination even a mechanism for producing an oscillating dipole can be found and hence the possibility of radiation under the appropriate circumstances. (An electron displaced from its equilibrium position will experience a net restoring force.)

Both these pictures are no more than models and one really cannot be believed to the exclusion of the other without some experimental verification of the model properties. An immediately apparent way to test these models is by the behavior of heavy charged particles when scattered by an atom. Clearly the major difference between the two models is the "size" of the heavy particles—protons—and this difference in size provides unambiguously different scattering behavior under the right conditions.

3.3 POTENTIAL SCATTERING   Take a very heavy atom, for example, gold. We will consider the scattering by it of a heavy positive charge, fully ionized helium. We will ignore the effects of the electrons, since they are so much lighter they will be knocked away with negligible energy loss. Also, by choosing helium, center of mass corrections in the scattering problem will be of little importance. We just end up with simple repulsive potential scattering between two charges. The form of the potential of the heavy scattering center obviously is going to be very different for the two atom models. The Rutherford model predicts point charge Coulomb scattering for all projectile-target separations and the Thomson model predicts point-charge Coulomb scattering *only* down to the atomic dimension, r_{atom}. As can be seen in Fig. 3-7, these two potentials predict a *very different* scattering behavior if we choose an initial kinetic energy, greater than the critical value KE_c, given by

$$KE_c = \frac{3}{2} \frac{kZ_1Z_2e^2}{r_{atom}}$$

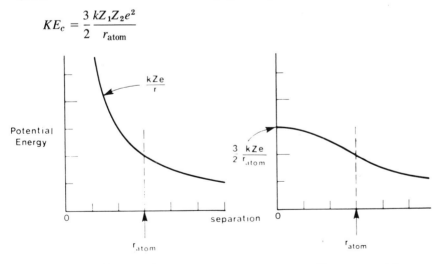

FIGURE 3-7 Scattering potentials for the Rutherford and Thomson charge distributions.

where Z_1 and Z_2 are the number of protons in the projectile and target, respectively.

Quite clearly the Rutherford model predicts backscattered particles (scattered through angles to greater than 90°) for such kinetic energies and the Thomson model predicts *only* forward scattering. For the case of gold the corresponding potential is

$$V = \frac{3}{2} \frac{kZ_{\text{gold}}e}{r_{\text{atom}}}$$

The size of atoms is known to be of the order of 10^{-10} m so that for incident helium ions with kinetic energies greater than about 3 keV, the presence or absence of backscatter will distinguish clearly between these two models.

Precisely this scattering experiment was done by Rutherford. He used 5.5 MeV helium atoms stripped of their electrons (α particles) to bombard gold atoms. At that energy, several thousand times KE_c, he saw backscattering! The Thomson model is therefore untenable.* In fact, from the results of his experiment Rutherford was able to conclude that the positive charge of the gold atom was confined to a dimension on the order of 10^{-14} m. The atom has a very compact central positive charge consistent only with the Rutherford model. The Rutherford experiment unambiguously points us at the orbiting picture of the atom, and gives us back the problem of radiation collapse!

3.4 WAVES AND PARTICLES   We have gone from the problem of understanding the particulate aspect of light to the problem of understanding how charged particles in accelerated motion contrive not to emit electromagnetic radiation. It is not too difficult to realize that there is a connection between the two problems, and so it should not be too surprising to find that we can gain some insight into the root of the cause of the latter problem by looking again at the former. This time we will use some new insight, provided by de Broglie.

It is clear that a beam of light transports energy. What may not be so clear, although equally true, is that associated with this energy transport is a *momentum* transport. If a beam of light is incident on a surface for some time so that its energy is absorbed by the material, then there is also a momentum p delivered to the object which, according to classical electromagnetism, is related to the delivered energy E by

$$p = \frac{E}{c}$$

Now de Broglie made use of this relation, and the photon energy relation $E = h\nu$, combining them to give

$$p = \frac{h\nu}{c} = \frac{h}{\lambda}$$

* For a detailed discussion of the interpretation of this experiment see Further Reading, Norwood, sec. 6-2.

This simple relation can be made to be quite revealing. If light consists of particles, as we have been forced to decide, and it has a wavelength (or wave property), as we have always believed, then the above is telling us the way in which the old and expected property (wavelength) is related to the new and unexpected particulate property (momentum). It tells us how to convert what we know, wavelength, into what is unexpected, the momentum of a photon.

We also can obtain this relation in a way that does not rely specifically on properties of electromagnetic waves. We can start with the general mass-energy relation

$$E = \sqrt{p^2 c^2 + m_0^2 c^4}$$

which is true for any object. One such object is a photon; it is really a particle in every respect except that it has no rest mass. Therefore for a photon the above relation becomes $E = pc$. This gives us

$$E = h\nu = pc$$
$$\frac{h\nu}{c} = p$$

where ν is the frequency of the light that the photons represent. For the photons, with corresponding wavelength λ, this takes on the same form as before, that is,

$$\frac{h}{\lambda} = p$$

and it is no longer clear whether this is a relation that should be applied to waves or to particles. De Broglie's decision, on considering this question, was that the relation is a general one. It relates both momentum to wavelength *and* wavelength to momentum. It says that waves will act like particles *and* particles will act like waves. It *predicts* that particles will act like waves, or rather de Broglie decided that it made this prediction.

If his prediction is true, it is necessary to decide under what circumstances the wave nature of a particle might become important. (Certainly the photo effect and light diffraction teaches us that different aspects of light are important in different circumstances.) Specifically, will the wave nature of an electron (whatever it may be) be significant on the size scale of an atom? What would the "wavelength" of an electron in a hydrogen atom be?

$$\lambda_{el} = \frac{h}{p} = \frac{h}{\sqrt{2 m_e KE}}$$

(Note that we use the nonrelativistic approximation for p.)

From the previous solution to the dynamics of the hydrogen atom this is

$$\lambda_{el} = \frac{h}{\sqrt{(m_e \, ke^2)/r_{atom}}} \cong \frac{6.6 \times 10^{-34}}{\sqrt{[10^{-30} \times 9 \times 10^9 \times (1.6 \times 10^{-19})^2]/10^{-10}}}$$

$$\lambda_{el} \cong 3 \times 10^{-10} \text{ m} = 3\,\text{Å}$$

where we have taken the radius of the hydrogen atom, r_{atom}, to be approximately 1 Å. The "wavelength" of the electron is *comparable* to the size of the atom when it has the energy necessary to orbit around the proton. The electron's wave nature is therefore most certain to play an important role in the properties of atoms.

We are now faced with the problem of incorporating a property that we do not understand (electron waves) into any model of the atom. Whatever is done at this stage will have to be simple enough not to require any basic understanding of the wave nature, yet basic enough to incorporate properly the essential property of a wave.

What is needed is a way of "looking" at what is going on in an atom that can be described both in terms of waves *and* particles. The basic property of the motion of the electron around the proton (nucleus) is that it stores energy. It also is possible to store energy in waves. If the energy of the electron wave is to be stored in the atom then it must be in the form of a *standing wave*. Any other mode will either transport the energy away, or cease to exist through eventual self-cancellation.

Clearly the simplest way to look at the problem is to incorporate a standing wave condition into the old dynamic treatment. We can not expect such a treatment to be proper, or even particularly well-informed about what is really happening, but it will at least take cognizance of our new awareness of the general wave-particles duality signaled by the relation $\lambda = h/p$.

3.5 THE BOHR ATOM

The following is the description of the atom due to Bohr, but using the basic reasoning of de Broglie, which came after Bohr's original work.* This treatment is therefore a composite, and takes advantage of hindsight.

For simplicity we first will assert that the "orbit" of an electron around the nucleus can be taken to be circular but the orbit length, or circumference of the circle, must be an integral number of "wavelengths." This is the *standing wave condition*.

$$n\lambda = 2\pi r$$

any integer value
radius of electron orbit

This condition must be incorporated into the previous energy equations for the motion of the atom (for simplicity we will assume a hydrogenlike atom of 1 electron circling a nucleus of Z protons).

$$TE = -\frac{k\,Z\,e^2}{2r} = -KE$$

* See Further Reading, Weidner and Sells, Alternate 2nd edition, sec. 6-4.

The standing wave condition is related to the electron's momentum so that we find

$$r = \frac{n\lambda}{2\pi} = \frac{nh}{2\pi \, m_e v} = \frac{n\hbar}{m_e v}$$

(The symbol $\hbar$ is used as a shorthand notation for the quantity $h/2\pi$). To use this relation in the total energy equation we must remove the dependence on electron velocity v. This can be done in the following way:

$$r^2 = \frac{n^2\hbar^2}{m_e^2 v^2}$$

but

$$m^2 v^2 = 2m(KE) = 2m\frac{kZe^2}{2r}$$

or

$$m_e^2 v^2 = \frac{m_e kZe^2}{r}$$

so that

$$r = \frac{n^2\hbar^2}{m_e kZe^2}$$

Now the radius can be seen to depend only on basic constants and may be inserted directly into the energy relation to obtain

$$TE = \frac{-kZe^2}{2r} = \frac{-kZe^2}{2} \times \frac{m_e kZe^2}{n^2\hbar^2}$$

$$TE = \frac{-m_e k^2 Z^2 e^4}{2 n^2 \hbar^2}$$

$$= \frac{-1}{n^2} Z^2 \left(\frac{m_e k^2 e^4}{2\hbar^2} \right) = \frac{-Z^2 E_0}{n^2}$$

$$\frac{m_e k^2 e^4}{2\hbar^2}$$

The inclusion of the standing wave condition puts the restriction on the total energy that *only certain values* of the total energy are possible, as illustrated in Fig. 3-8. All other energies would give a combination of wavelength and radius which would cause an eventual self-cancellation of the electron wave, and it could not exist permanently in that orbit. Most importantly we see that there is a lowest possible, or most tightly bound state, below which the motion of the electron can never satisfy the standing wave condition, and therefore cannot exist. The atom must be stable against further energy loss—radiation—when it reaches the $n = 1$ state. If we ask why the electron does not radiate at that state we can only avoid the question at this point by remem-

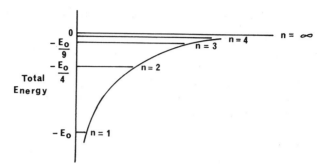

FIGURE 3-8 Allowed values of total energy for an electron in a hydrogen atom when the standing wave requirement is accounted for. E_0 is the ground state binding energy, and n is the quantum number.

bering that we are dealing with a wave now, and that this has modified the description radically. An answer to "why" about the nonradiation will have to wait for more understanding of what the wave is.

This Bohr picture has introduced several new things. As we saw, the energy is restricted to discrete values, specified by the integer n. The energies are said to be *quantized* and the integer n is called a *quantum number*. For small values of n the quantum property is important—the allowed energies are well separated; for large values of n—where the electron becomes increasingly more weakly bound—the difference in allowed energies becomes smaller, to the point of insignificance. For the very high values of n the quantum condition effectively ceases to put any restrictions on the "allowed" energies and hence the motion of the electron; that is, we return to classical physics. This return to classical behavior when quantum numbers become very large must be true for any quantum effect. In its general form this is called the *correspondence principle*.

Energy is not the only property that has become quantized. There are several others, and in particular we can see from the standing wave condition

$$n\lambda = 2\pi r$$

that as a direct consequence of the wave nature of the orbiting electron

$$n\hbar = mvr = L$$

the orbital angular momentum L has become quantized. In the Bohr picture this is really no different from saying that the energy is quantized, but later this will take on an independent significance.

From the discrete nature of the possible states, and from the fact that the motion of the electron is such that it can not have energies in between these allowed states, we can see how light radiation can occur not continuously, but as discrete entities. The electron, when in an energy state higher than the most tightly bound one, can be thought of as continuing to experience an attraction toward the central positive charge, which "pulls" it into the more tightly bound state. However, since the electron cannot have a permanent state of motion for energies between the two allowed ones, it must make the transition by an "instantaneous jump" from one energy state to another. To conserve total energy the system must emit radiation, and it must be done between the times the electron is in one or the other states of allowed motion. This should happen within a few cycles of the electron's orbit, at most. A more detailed picture of how the transition occurs (which is actually beyond the scope of the Bohr model) can be obtained by considering the two standing wave states involved in the "jump" to be analogous to two resonant cavities that are (weakly) coupled together. Initially we can imagine the electron to be entirely in one of the standing wave states and not radiating. However, as the electron wave "leaks" into the lower state, the pure standing wave condition is destroyed and radiation is emitted which carries off energy so that the electron is obliged to exist in the lower state.*

While it is essential that a model (the Bohr atom) provide a correct qualitative picture of the behavior of a system, it really only becomes useful (and strictly testable) when quantitative results are considered. There are several such predictions possible with the Bohr model, and we will look at them now.

The electron in a hydrogen atom is predicted to end in a stable configuration with a total energy given by the quantum number $n = 1$

$$TE = -\frac{E_0}{(1)^2}$$

It requires this amount of energy to remove the electron from the atom (this is called the ionization potential). Quantitatively this energy amounts to

$$E_0 = \frac{m_e k^2 e^4}{2\hbar^2}$$

$$= \frac{9.110 \times 10^{-30} \times (8.987 \times 10^9)^2 \times (1.6022 \times 10^{-19})^4}{2 \times (1.054 \times 10^{-34})^2}$$

$$= 218.02 \times 10^{-20} \text{ J}$$

$$= \frac{218.02 \times 10^{-20} \text{ J}}{1.6022 \times 10^{-19} \text{ J/eV}}$$

$$= 13.60 \text{ eV}$$

The electron is predicted to be bound to the hydrogen nucleus by E_0, or 13.60eV. The observed value is 13.58 eV, in excellent agreement!

The Bohr atom not only predicts the binding energy, but also energy differences of allowed states. These energy differences should correspond to observed energies of

* For further discussion of the decay process, see Further Reading, Beiser, sec. 6-9.

radiation of the hydrogen atom, as the electron radiates down to the stable ground state.

In general, the possible energy differences are

$$\Delta E_{\text{allowed}} = E_{n_1} - E_{n_2}$$

$$= \frac{E_0}{n_1^2} - \frac{E_0}{n_2^2}$$

$$= E_0 \left(\frac{1}{n_1^2} - \frac{1}{n_2^2} \right)$$

This energy is related to the frequency of the emitted radiation by $\Delta E_{\text{allowed}} = h\nu_{\text{emitted}}$

$$h\nu = \Delta E = E_0 \left(\frac{1}{n_1^2} - \frac{1}{n_2^2} \right)$$

$$\nu = \frac{E_0}{h} \left(\frac{1}{n_1^2} - \frac{1}{n_2^2} \right)$$

where n_1 and n_2 are any integers. The experimentally observed radiation frequencies for the hydrogen atom were found to fit the formula

$$\nu_{\text{exp}} = cR \left(\frac{1}{n_1^2} - \frac{1}{n_2^2} \right)$$

where R is a constant. This relation has exactly the same dependence on the integers n_1 and n_2 that is predicted by the Bohr model. All that remains is to compare the value of E_0/h with cR. R, the so-called Rydberg constant, was determined by fitting the spectral lines of hydrogen to be

$$R = 109677.8 \ cm^{-1}$$

so that $cR = 2.997924 \times 10^{10} \times 109677.8 = 328.8 \times 10^{13}$ Hz. This should be compared with

$$\frac{E_0}{h} = \frac{13.6 \ \text{eV} \times 1.6 \times 10^{-19} \ \text{J/eV}}{6.6 \times 10^{-34} \ \text{J s}} = 329 \times 10^{13} \ \text{Hz}$$

which proves to be in extremely good agreement.

3.6 NONHYDROGEN BOHR ATOM   In deriving the Bohr atom formulae we allowed for the possibility of more than one proton in the nucleus, but only one electron. But the model itself, through quantized radii, suggests that the orbiting electrons in any atom will be distributed *in shells* about the nucleus, as illustrated in Fig. 3-9. For this reason we can extend the Bohr atom predictions to describe certain

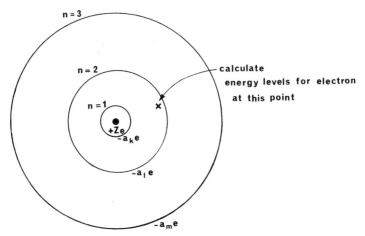

FIGURE 3-9 Bohr model for multielectron atoms. Orbits are labeled by the quantum number $n = 1, 2, 3$, and so on, and the corresponding number of electrons by a_k, a_l, and a_m.

de-excitations for a range of atoms using the following properties of charge distribution.

For a uniform spherical shell charge distribution, if we are *outside* the sphere, we can treat the distribution as if it were a point charge at the center of the sphere. If we are *inside* the shell there is no net field; we will feel no net force due to the distribution. This means that if we choose a particular electron in one of the shells, we can calculate the energy levels for it by ignoring all electrons that are further from the nucleus than itself, and by reducing the nuclear charge Z by the number of electrons in shells closer to the nucleus than itself. This gives the result

$$E_n = \frac{Z_{\text{eff}}^2 E_0}{n^2}$$

$$= \frac{(Z - a)^2 E_0}{n^2}$$

where a is the number of electrons "shielding" the nucleus. We will temporarily ignore the problem of other electrons in the same shell as the electron undergoing a transition. We cannot treat the problem of other electrons in the same shell in this simple way, and ultimately will have to be guided by observation in deciding how to treat them. In order not to press the picture too hard (which it could not stand) we will only consider levels that differ by 1 in their quantum number. In this case a is taken to be the same value for both levels. We can not predict from our model what a should be for any particular n value or whether it should be the same for different elements, but anticipate that a will have a well-defined value for any choice of n and that it will be independent of the element.

This idea can be checked in the following way. Assume a material is bombarded with high-energy particles, say the electrons from a scanning electron microscope. If these bombarding particles are energetic enough, they can eject an electron in one of the inner shells. There will be a subsequent de-excitation of the electron structure to fill the vacancy created in the inner shell, as shown in Fig. 3-10.

Of all the various radiations given off in the subsequent radiations, we will consider here only the case where an electron from the next shell fills a vacancy created in the $n = 1$ shell (this type of transition is labeled a K_α transition).

For this transition we can write the transition energy

$$\Delta E = (Z - a_k)^2 E_0 \left(\frac{1}{1^2} - \frac{1}{2^2} \right)$$

$$= \frac{3}{4} (Z - a_k)^2 E_0$$

where a_k is the number of electrons in the innermost shell minus one (this innermost shell is labeled K), as there is a vacancy when the transitions occur. Notice also that we are assuming that a is the same for the electron before and after the transition. This is equivalent to saying that the electron that makes the actual transitions does so from just slightly "inside" its shell. Therefore if we plot $\sqrt{\Delta E}$ vs Z for $n = 2 \to n = 1$ transitions in various materials we should get a straight-line plot (called a Moseley plot) with a slope of $\sqrt{0.75 E_0}$, and an intercept at a. If we look at the experimental results, shown in Fig. 3-11, we see that the agreement is indeed very good; the actual slope is $\sqrt{0.76 E}$, and the intercept is at $a = 1$, indicating that there are two electrons in the K shell. The same comparison can be carried out for the analogous transition to the $n = 2$ shell from the $n = 3$ shell (L_α transition); in this case the slope should be $\sqrt{0.139 E}$, and the observed slope is $\sqrt{0.136 E_0}$. The intercept is at $a = 9$, indicating that there are eight electrons in the $n = 2$ shell.

The good agreement between observation and calculation gives us confidence in the use of the same a for initial and final states. The interpretation of a in terms of numbers of electrons in shells will be found to be correct in sec. 4-6, so that we can use the systematic information, as shown in Fig. 3-11 to guide our choice of a in various circumstances.

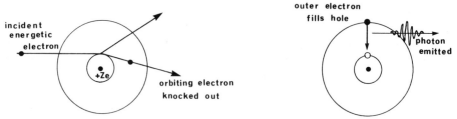

FIGURE 3-10 Ejection of atomic electron, resulting in photon emission.

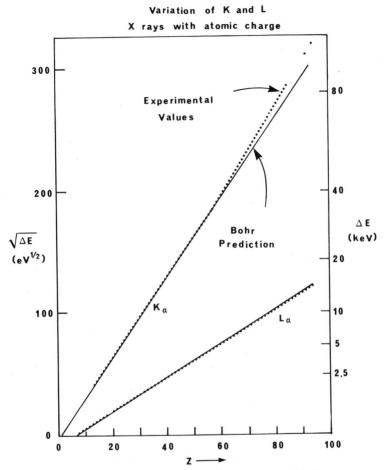

FIGURE 3-11 Variation of K and L X-ray energy with element.

There are a great many transitions that can be identified in the systematic way suggested in Fig. 3-11. The general scheme of nomenclature for these transitions is illustrated in Fig. 3-12; a transition is labeled by the level to which the electron de-excites using a scheme where K corresponds to $n = 1$, L corresponds to $n = 2$, M to $n = 3$, and so on. The level from which the electron de-excites is identified by subscripting the transition in order of increasing energy with Greek letters. Hence an $n = 2 \rightarrow n = 1$ transition is a K_α transition; $n = 3 \rightarrow n = 1 = K_\beta$ and $n = 4 \rightarrow n = 2 = L_\beta$, and so on. For any set of transitions to a particular shell, the one from the adjacent quantum level (i.e., the α-subscripted transition) is the most probable one to occur, so that K_α, L_α, and so on, transitions are the most intense ones normally observed. This corresponds to the fact that electrons in the shell just outside a vacancy

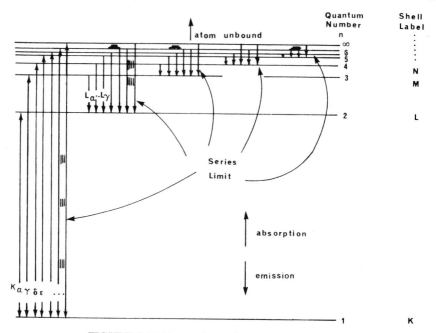

FIGURE 3-12 Nomenclature for atomic transitions.

are most strongly "attracted" to it, since there is a minimum amount of intervening shielding charge.

All these transitions can be readily identified experimentally from the systematics of their energies, and consequently the element giving rise to them can be identified uniquely, even using the simple Bohr model. Notice that the energies involved in these transitions are quite high, in the keV range; they are generally called characteristic X rays. These characteristic X rays serve as a very useful identification "tag" for elemental analysis in such devices as scanning electron microscopes, and were used to provide the information on elemental distributions in the SEM picture shown previously. The characteristic X rays, due to transitions of the innermost electrons, are particularly useful in elemental analysis since, as our treatment of the many-electron Bohr atom shows, the innermost electrons are virtually independent of any properties of the outer electrons, which can be altered by details of bonding to other atoms. Therefore, K X-ray energies are reliable indicators for the elements, independent of their chemical form.

3.7 THE BOHR MAGNETON The Bohr model also makes some definite predictions about magnetic properties of the atom that are worth investigating. The picture of an electron circulating around the nucleus is equivalent to a current loop. From our knowledge of the relation of currents to magnetic fields (Ampere's law) we know that the permanent current loop predicted by the Bohr picture should give rise to a permanent magnetic field, which will be in the form of a dipole field.

The field generated is described in terms of a magnetic dipole moment M

$$\underset{\text{dipole}}{\mathbf{B}(r)} = \frac{\mu_0}{4\pi}\left(-\frac{M}{r^3} + \frac{3M \cdot \mathbf{r}\,\mathbf{r}}{r^5}\right)$$

where (Fig. 3-13) $M = i\mathbf{A}$

and A is the area enclosed by the current loop that is carrying a current i.

This is exactly analogous to an electric field due to an electric dipole moment. What is of interest is the fact that the electron motion should create a magentic dipole moment (which gives rise to a magnetic field) that can be calculated precisely using the Bohr model.

The magnetic moment for an electron circulating in a Bohr orbit is

$$M = \pi r_n^2\, i$$

where r_n is the radius of the nth quantum level and

$i = 1$ electron per cycle

The time T required for an electron to complete a cycle is given by the ratio of the circumference of the circle to the velocity v_n of the electron

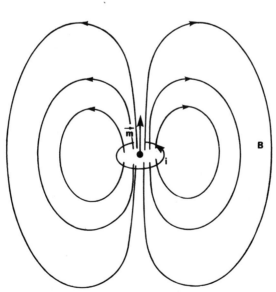

FIGURE 3-13 Magnetic moment of a current loop and associated magnetic field.

$$T = \frac{2\pi \, r_n}{v_n}$$

hence

$$i = \frac{e \, v_n}{2\pi \, r_n}$$

so that

$$M = \frac{\pi r_n^2 \, e \, v_n}{2\pi \, r_n}$$

$$= \frac{\pi \, r_n \, v e}{2\pi} = r_n m_e \, v_n \, \frac{e}{2m_e}$$

However the quantity $m_e r_n v_n$ is just the electron's angular momentum, which we have already worked out

$$L = n\hbar$$

so that

$$M = \frac{e}{2m_e} L = n \frac{e\hbar}{2m_e}$$

We find that the Bohr model predicts that the magnetic moment of an atom is quantized, in multiples of $e\hbar/2m_e$. This quantity is very useful in discussing magnetism on the atomic scale and is called the Bohr magneton, denoted by μ_B.

It is now appropriate to "abandon" the Bohr model. From the beginning we have known that it was only the simplest possible picture that could be made. Historically, there were several refinements of this model; inclusion of center-of-mass corrections, elliptical orbit calculations and relativistic corrections, but none questioned the basic correctness of accepting a model based on only the most gross wave property possible. Any real progress can only come with a more basic understanding of this new aspect of matter.

FURTHER READING   A. Beiser, *Concepts of Modern Physics,* 2nd ed., McGraw-Hill, 1973.

J. Norwood, *Twentieth Century Physics,* Prentice Hall, 1976.

L. Wegman, The Photo-emission Electron Microscope: Its Technique and Applications, *Journal of Microscopy* 96, (1972) 1.

R. T. Weidner and R. L. Sells, *Elementary Modern Physics,* Allyn & Bacon, 1973.

PROBLEMS   1. A new photoemitting material is being tested. Irradiation by a sodium lamp ($\lambda = 5890$ Å) produces electrons with a maximum kinetic energy of 0.380 eV. What is the work function of this material? What is the longest wavelength light that will produce the photoelectric effect in this material? What is the maximum kinetic energy electrons that can be produced with a mercury arc lamp ($\lambda = 2537$ Å)?

2. Calculate the threshold frequency for photoemission in Cs, K, and Sr. Roughly to what colors do these frequencies correspond? Describe how you might be able to design a color-sensing device using these metals.

3. The threshold frequencies for photoemission in K and Zn are 0.38 and 1.03×10^{15} Hz, respectively. What are their work functions?

4. Explain why the photoelectric effect shows that light is quantized.

5. What would the equilibrium separation of the two electrons in a helium atom be according to the Thomson model? Assume an atom radius of 1Å(1×10^{-10} m).

6. Prove that the potential at the center of a uniformly charged sphere with total charge Q_0 and radius r_0 is 3/2($k\ Q_0/r_0$). Calculate the force on a charge q as a function of the distance from the center of Q_0.

7. What is the de Broglie wavelength of a 20 keV electron? How does it compare with the de Broglie wavelength of a proton and a photon of the same energy? Is the wave property of the electron important for electron microscopes? Why?

8. Using the Bohr model, predict the value of the ground state and first two excited state energies of He$^+$ (helium with a net single positive charge). Do you expect these predictions to be reasonably accurate? Why?

9. List the elements whose excitation energies should be closest in value to the Bohr predictions, stating the reason for your choice.

10. Using the standing wave arguments developed for the Bohr atom, determine what the energy level structure would be for an electron confined to a one-dimensional box 1Å in length.

11. What is the classical radiation pattern for a closed current loop? Compare this to what is observed in the hydrogen ground state.

12. The temperature of an incandescent gas of atomic hydrogen (a plasma) is to be measured remotely. The intensities of two transitions known to be the $n = 2$ state are in the ratio of 5/1; their frequencies are 6.2×10^{14} Hz and 4.5×10^{14} Hz. What is the temperature of the gas? (Assume that the radiation is entirely due to conversion of kinetic energy of a colliding atom to the various excited states. In this energy region the Maxwell-Boltman distribution can be taken as proportional to $e^{-E/kT}$).

13. Calculate the K_α X-ray energies of Na, Cr, and Pb using the Bohr model.

14. In attempting to identify elements by characteristic X-ray emission, what particular set of X rays (e.g., L_β, M_α) will provide the greatest energy separation between emission from elements that differ by 1 in Z?

15. A material being investigated by a scanning electron microscope has in it Mn(Z = 25) and Fe(Z = 26). Approximately what energy resolution would be required of an X-ray detector in order to be able to distinguish between K_α X rays of the two elements? If the electron beam energy is 7 keV, would you be able to excite these X rays? Why?

16. A group of elements is being studied by induced X-ray analysis, where the X rays are produced by bombarding the elements with high-energy particles. All of the highest-energy X rays have been identified down to an energy of 24.490 keV, where an as-yet unidentified line appears. What element is this line due to? What is the value of the highest energy that you could look for the existence of another line to confirm your identification?

17. Calculate the value of the Bohr magneton in eV/Tesla.

18. A muon is a particle that is very similar to the electron except that it is approximately 207 times more massive than the electron. If a muon were to become bound to a proton, what would you expect its binding energy to be? How large would the "atom" be?

19. What would the magnitude of a Bohr magneton be for the muon?

FOUR

THE QUANTUM ATOM

The qualitative and quantitative successes of the Bohr model for hydrogen and hydrogenlike atoms strongly suggests that we must accept the reality of the de Broglie wave aspect of matter. However to proceed any further we must attempt to determine what the "wave" is. To do this we should first examine a property of all physical waves—how well we can know their frequencies.

4.1 DE BROGLIE WAVES AND THE UNCERTAINTY PRINCIPLE If we wish to know what the frequency ν of some wave is, we would measure it, essentially by counting the crests or troughs of the wave that pass by us in some definite time interval, that is,

$$\nu = \frac{N}{\Delta t}$$

where N is the number of crests we count and Δt is the duration of our measurement. However we must recognize that this formula is exactly correct *only* if the wave is a completely pure sine wave so that the amplitude does not vary. That is, it must be a wave with only one frequency; otherwise we might well miss cycles due to beats. Such a situation (pure single-frequency waves) is completely unphysical since real waves must start and stop at some time; this very fact guarantees that there are other frequencies present (from Fourier analysis of waves). To get an idea of how this will affect our knowledge of the properties of a physical wave train we can consider the following. In the simplest case we could have a wave consisting of no less than two components with very nearly equal frequencies so that at some time during our measurement a beat might have occurred and one of the crests (or troughs) was missed. We must therefore recognize that our measurement of crests is indeterminate by the amount we may (or must) have missed during our measurement period so that there is an *indeterminacy* $\Delta\nu$ of the measured frequency given by

$$\Delta\nu \geqslant \frac{1}{\Delta t}$$

We can easily reduce this indeterminacy by measuring for a longer time, but there will always exist the relation

$$\Delta\nu \, \Delta t \geqslant 1$$

Such a relation can be obtained more elegantly by looking at the properties of a Fourier integral representation of a finite-length wave train, but the fundamental conclusion remains the same. The longer you take to make a frequency measurement, the more accurately you can do it. (A proper Fourier analysis, however, reduces the uncertainty by a factor 4π, basically due to the fact that the waves can be measured more accurately than by counting crests. We will take over the factor of 4π in our treatment.) This relation must also be true for light and specifically the photon. Hence, beginning with

$$\Delta\nu \, \Delta t > \frac{1}{4\pi}$$

and using $E = h\nu$, we see that there must be some uncertainty in the energy

$$\Delta E = h\Delta\nu$$

related to the length of time that it takes to measure the photon energy. This gives

$$\Delta E \, \Delta t > \frac{\hbar}{2}$$

This can also be rewritten in a form more appropriate to particles as

$$\frac{\Delta E}{c} \, c\Delta t > \frac{\hbar}{2}$$

The first term is the photon momentum indeterminacy (uncertainty) and the second term is the indeterminacy (uncertainty) of the photon's position

$$\Delta p \, \Delta x > \frac{\hbar}{2}$$

(Δp refers to the component of p in the x direction; we would find that this is so if we were to make a proper derivation.) This relation between uncertainties is called the *Heisenberg uncertainty principle*.

We have used the photon to reformulate a basic property of finite wave trains into "particle language." Now we must interpret them to learn their physical meaning. Let us begin with the classical property of wave diffraction.

If we take a beam of light and pass it through a single slit (Fig. 4-1) we would get a widely spread (diffracted) pattern of light if the slit is comparable to a wavelength, or just an image of the slit if the slit is much larger than a wave of light. This can also be interpreted, using $p = h/\lambda$, from the point of view of a stream of photons rather than plane waves of light. If we constrain the photons to pass through the slit opening then we know their position at the slit to within an uncertainty Δx.

Therefore, we cannot know its momentum in the x direction to better than Δp, where $\Delta p = \dfrac{\hbar}{2\Delta x}$. The bigger the slit the better defined the momentum, and the less the beam spreads; the smaller the slit the more poorly defined the momentum, and the more the beam spreads.

Both descriptions of the light give the same result; the wave picture gives the intensity or amplitude distribution of the wave at the screen as a function of the ratio of the slit dimension to the wavelength. The particle picture gives us the concept of a distribution of relative likelihood, or probability, that a photon will strike the screen in some region, which is determined by the slit dimension and the uncertainty in momentum. The wave amplitude performs the same function as the likelihood distribution.

Now, because of the de Broglie hypothesis and our acceptance of it, we must come to the conclusion that if we did the *same* thing with electrons the same relations would have to apply.

As illustrated in Fig. 4-2, after passing through the slit of width Δx these electrons have a transverse momentum that is not fixed to within $\Delta p > \hbar/2\Delta x$. This means we *cannot* predict *exactly* where individual electrons will strike the screen. We can only give the relative probability that they will strike different parts of the screen.

When a large number of electrons have passed through the slit there will be a pattern of the number of strikes vs position on the screen, analogous to the intensity of light

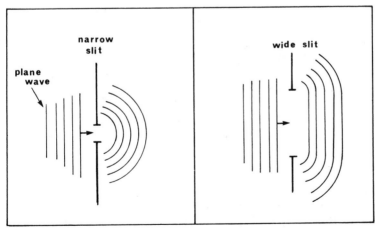

FIGURE 4-1 Single-slit diffraction of waves for narrow and wide slits.

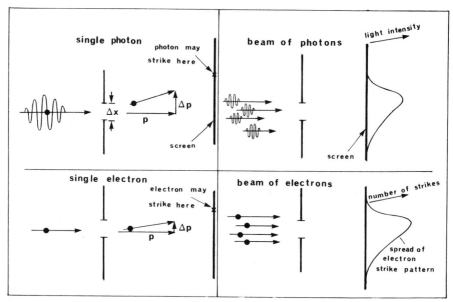

FIGURE 4-2 Single-slit diffraction of photons and electrons.

on the screen. For a single electron (or any particle, for that matter) the pattern will represent the probability of finding that electron at any particular point on the screen. Just as the detailed intensity pattern for light is determined by the wavelength, so the detailed probability pattern is controlled by the wavelength of the electron. *The wave nature of matter is what determines the probability of finding the particle at a particular point in space and time.*

For light there are two properties that can be described by the wave; one is the light amplitude, the other is the light intensity, which is the square of the amplitude. We must decide which of the two we are dealing with in particle waves. This can be decided by the following example of two-slit interference illustrated in Fig. 4-3. If the particle wave is a ''probability wave,'' (i.e., the wave itself gives the probability of finding the particle and is therefore analogous to light intensity) then when the two slits are used the total pattern on the screen will give the sum of the patterns that would be produced by each slit in isolation; there would never be any cancellation between the two patterns. On the other hand, if the wave is a probability *amplitude* (i.e., the wave must be *squared* to obtain the probability of finding the particle), then there can be cancellation between the two waves, with the possibility of nulls in the pattern of strikes, which could not be produced in the other case.

This experiment actually can be performed with particles (electrons) and the result is that cancellation is observed. Consequently the wave must be a probability amplitude wave. The de Broglie hypothesis finally leads us to the conclusion that associated with particles there is a wave such that the square of that wave at any given point in

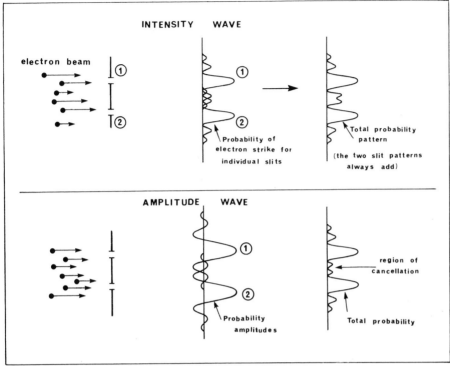

FIGURE 4-3 Two-slit interference patterns for amplitude and intensity waves.

space and time gives the probability of the particle being found at this position. All particles are now described by a *wave function* $\psi(x,t)$ such that $[\psi(x,t)]^2$ gives* the probability that the particle is within dx of the point x at time t. All the information we can know about the particle is contained in this probability amplitude wave.

4.2 THE MECHANICS OF WAVES   We now realize that the maximum information that can be known about any particle is limited, and given by the probability amplitude wave. In order to properly treat the behavior of particles we must allow for the limitation to what can be known. By using this wave function, or distributed probability, rather than the classical notion of perfectly defined point objects, we should now be able to treat particle dynamics properly.

We can see how to do this from the following: take a general traveling wave y, in one dimension with an amplitude A, period T, and wavelength λ

---

* The absolute square has been used to ensure that the probability is positive under all circumstances.

$$y(x,t) = A \sin 2\pi \left(\frac{t}{T} - \frac{x}{\lambda} \right)$$

From the properties of sinusoidal functions we know that

$$\frac{\partial^2 y}{\partial x^2} = \frac{-4\pi^2}{\lambda^2} A \sin 2\pi \left(\frac{t}{T} - \frac{x}{\lambda} \right)$$

$$= \frac{-4\pi^2}{\lambda^2} y$$

This must apply equally well to our matter wave ψ

$$\frac{\partial^2 \psi}{\partial x^2} = \frac{-4\pi^2}{\lambda^2} \psi$$

However, since the wavelength of the matter wave is related to the particle momentum by

$$\frac{h}{p} = \lambda$$

this becomes

$$\frac{\partial^2 \psi}{\partial x^2} = \frac{-4\pi^2}{h^2} p^2 \psi$$

$$= \frac{-p^2}{\hbar^2} \psi$$

For a nonrelativistic particle the momentum is simply related to the kinetic energy so that

$$\frac{\partial^2 \psi}{\partial x^2} = \frac{-2m \, KE}{\hbar^2} \psi$$

If this particle is in a potential $V(x)$, then the above relation becomes

$$\frac{\partial^2 \psi}{\partial x^2} = \frac{-2m}{\hbar^2} [TE - V(x)] \psi$$

where TE is the total energy and $V(x)$ is the potential energy. This relation determines the distribution of the wave function for the particle (in one dimension) for a given potential energy distribution. It is called the Schrödinger equation, and is a second order differential equation which can be solved to yield a particular function $\psi(x,t)$ that fixes everything that can be known about the spatial distribution of the particle independent of time. The particular equation we have arrived at is a one-dimensional, time-independent Schrödinger equation; generally it is three-dimensional and time-dependent.

We could use the three-dimensional form of this equation to solve the problem of the hydrogen atom. We would have to insert the proper Coulomb potential for the V term, and rewrite the three-dimensional problem in spherical coordinates. The equation would then look like the following

$$\frac{1}{r^2}\frac{\partial}{\partial r}\left(r^2\frac{\partial\psi}{\partial r}\right) + \frac{1}{r^2}\sin\theta\frac{\partial}{\partial\theta}\left(\sin\theta\frac{\partial\psi}{\partial\theta}\right)$$

$$+ \frac{1}{r^2}\sin^2\theta\frac{\partial^2\psi}{\partial\phi^2} + \frac{2M}{\hbar^2}\left(TE - \frac{kZe^2}{r}\right)\psi = 0$$

for the single-electron hydrogenlike atom. Clearly the problem has become much more complex to solve than that of the Bohr atom. Its solution, however (the wave function for the electron) gives us *all* the information needed (or knowable) about the electron; the probability of finding the electron at a particular place, its energy, its angular momentum when in a particular total energy state.

The general form of the solution (which we will not derive as it can be found in standard quantum mechanics texts*) looks like

$$\psi = R_{n,\ell}(r)\Theta_{\ell,m_\ell}(\theta)\,\Phi(\phi)_{m_\ell}$$

where R is a function of r, labelled by two integer variables n and ℓ, so that R varies with r, the distance from the nucleus. This is called the radial wave function. Θ is a function of θ, labeled by two integer variables ℓ and m_ℓ, and $\Phi_m(\phi)$ is an exponential function of ϕ, labeled by the integer variable m_ℓ. We will first look at the integer variables, and then the general significance of the three functions.

The integer n is called the principle quantum number and plays the same role as the analogous quantity in the Bohr atom; it determines the total energy. In fact the solution for the energy levels of the hydrogen atom using the Schrödinger wave equations turns out to be

$$E_n = \frac{-mk^2Z^2e^4}{2\,n^2\hbar^2} = \frac{E_0}{n^2}$$

which is exactly the same as the Bohr solution!

The integer ℓ can have any value between 0 and $n-1$. This is due to the mathematical form of the $R_{n\ell}$ polynomial; the value of ℓ determines the angular momentum of the system. The actual magnitude of the angular momentum is

$$L = \sqrt{\ell(\ell+1)}\,\hbar$$

It is interesting to compare this with the Bohr predictions. The ground state of hydrogen has $n = 1$, and therefore ℓ can only be 0, and therefore $L = 0$. This is

* See Further Reading, Norwood, Ch. 9.

considerably different from the (incorrect) Bohr prediction of $1\hbar$. For large values of n, the maximum allowed angular momentum approaches the Bohr value

$$L_{\max} = \sqrt{\ell_{\max}\,(\ell_{\max} + 1)}\,\hbar = \sqrt{(n-1)n}\,\hbar \cong n\hbar$$

but there are also, for that same large n, many additional values of angular momentum not predicted by the Bohr model.

The quantity m_ℓ can be any integer from $-\ell$ to $+\ell$, and largely controls the nature of the ϕ- and m-dependence of the wave function. It is easiest to picture its role by thinking of L as a vector; then, as illustrated in Fig. 4-4, $m_\ell\hbar$ corresponds to the projection of that vector onto the z axis.

These projections, because of the nature of the solution of the Schrödinger equation, can only have integral values. Not only is the magnitude of the angular momentum quantized but its possible orientations are quantized. Basically this additional restriction can be recognized as a consequence of the fact that we are dealing with a three-dimensional wave, which must be a standing wave not only in the ϕ direction along the circumference (as was accounted for in the Bohr model) but also in the θ direction (over the top!).

Now let us look at the form of $R_{n\ell}$ and Θ_{ℓ,n_ℓ}. The radial wave function determines the probability of finding the electron at some distance r from the nucleus. This is shown for several values of ℓ and n in Fig. 4-5. If we look at the most probable distance for finding an electron in the $n = 1$ state, we find that it is at just the radius predicted by Bohr. The same is nearly so for the $n = 2$ orbit, although the radial position is significantly spread out. In addition the position distribution depends somewhat on the value of ℓ. The previous picture of concentric "shells" of electrons obviously has become at best a "fuzzy" one. There is considerable radial spread of the electron wave.

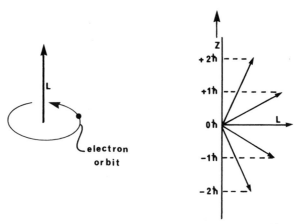

FIGURE 4-4 Quantized orientations of angular momentum. The case for $l = 2$ is shown.

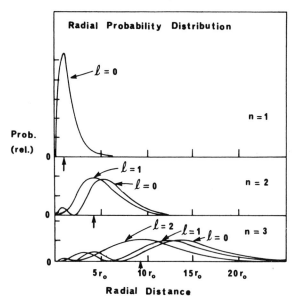

FIGURE 4-5 The probability of finding an electron within dr of r for various values of n and l. The radial distance is given in units of r_0, the Bohr value for the $n = 1$ orbit. The arrows indicate the Bohr prediction for the orbit radius. The quantity $r^2 R_{nl}$ has been plotted.

The function $\Theta_{\ell,m_\ell}(\theta)$ is particularly interesting. The θ-dependence of the probability distributions, several of which are shown in Fig. 4-6, shows that the electronic charge can be nonuniformly distributed, and that significant "local" concentrations of charge can occur. A simple $e^{im\phi}$ behavior is found for $\Phi_m(\phi)$. Linear combinations of $\Phi_m(\phi)$ and $\Phi_{-m}(\phi)$ lead to electron distributions shown in Fig. 4-7; from these distributions it is not hard to imagine how chemical bonding between atoms could have strongly directional character for the right kinds of electron wave functions. In fact it is the nature of these wave functions that plays a very large role in determining the "chemistry" of atoms.

The addition of the new quantum numbers means that the level structure of hydrogen has become more complex. For any value of n there is a multiplicity of possible ℓ values. This is accounted for by redrawing the energy level diagram to explicitly include the new quantum numbers.

It should not take much convincing to decide that for hydrogen it is meaningless to include m_ℓ since there can be no physical significance of the orientation of the electron orbit of a single electron orbiting a single proton. Any direction is exactly the same as any other direction. The energy of the atom does not change by reorienting an isolated hydrogen atom; the states with different m_ℓ values are said to be *degenerate*. (This is not necessarily true if there are several electrons present in an atom.)

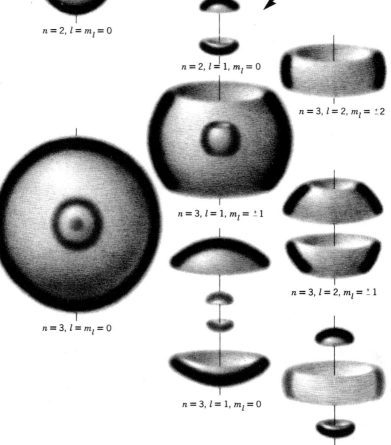

$n = 1, l = m_l = 0$

$n = 2, l = 1, m_l = \pm 1$

$n = 2, l = m_l = 0$

$n = 2, l = 1, m_l = 0$

$n = 3, l = 2, m_l = \pm 2$

$n = 3, l = 1, m_l = \pm 1$

$n = 3, l = 2, m_l = \pm 1$

$n = 3, l = m_l = 0$

$n = 3, l = 1, m_l = 0$

$n = 3, l = 2, m_l = 0$

FIGURE 4-6 Three-dimensional appearance of electron probability density for various values of n and l. The degree of shading indicates the relative probability of finding the electron. From *Quantum Physics of Atoms, Molecules, Solids and Particles* by R. Eisberg, and R. Resnick, Copyright © 1974, John Wiley and Sons, Inc. Reprinted by permission of John Wiley and Sons, Inc.

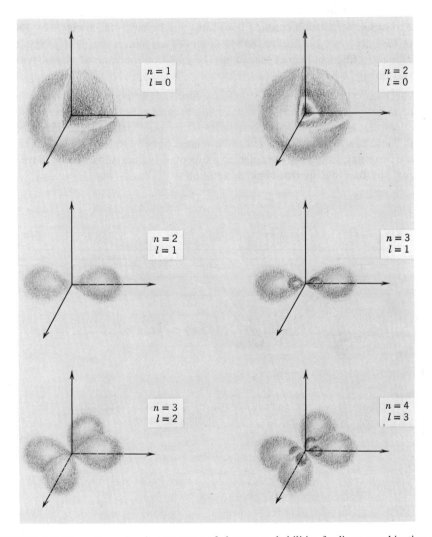

FIGURE 4-7 Three-dimensional appearance of electron probabilities for linear combination of $+m$ and $-m$ solutions for various values of n and l. The degree of shading indicates the relative probability of finding the electron.

The angular momentum, however, plays a significant role. Particularly, it should be recognized that photons that are emitted in a transition from one level to another must carry definite angular momentum, since angular momentum as well as energy is conserved, and the system has definite value of angular momentum both before and after the transition. A multipole expansion of an electromagnetic field radiating from a source shows (and we will take as given) that the most probable form of radiation field (the most intense multipole) is one that carries a unit total angular momentum. Hence most likely radiation-emitting transitions will be ones that change ℓ by 1. They are not the *only* type, but are by several orders of magnitude the most probable and hence the most frequent type of transition. The emission spectrum of a hydrogen atom is shown in Fig. 4-8.

It had been observed experimentally (before a formal description of the quantum atom) that the radiation could be classified according to their optical characteristics, such as sharp lines (S), lines of particular brightness (P for principal), diffuse lines (D), and so on. These characteristics turn out to be related directly to the angular momentum of the de-exciting state, and the angular momentum values subsequently have been "tagged" by their old spectroscopic description

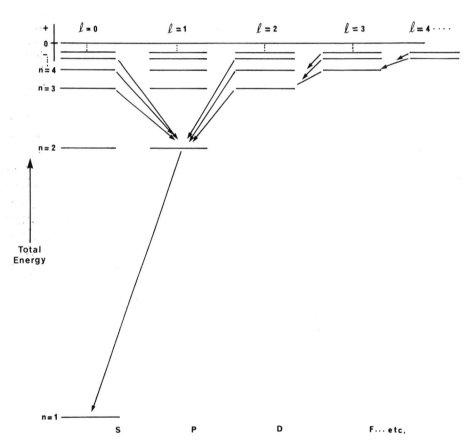

FIGURE 4-8 Quantum levels of atomic hydrogen, with their spectroscopic labels and associ-ated electromagnetic de-excitation transitions.

$$\ell = 0 <=> S$$
$$\ell = 1 <=> P$$
$$\ell = 2 <=> D$$
$$\ell = 3 <=> F$$
$$\ell = 4 <=> G$$

up to $\ell = 3$, beyond which the convention has simply extended the labeling alphabetically with increasing ℓ value.

4.3 TRANSITION RATES; RADIATION AND ABSORPTION   Once we know the detailed angular momentum properties of the various states of an atom, and the general charge distribution (as given by the wave function) we are in principle able to calculate the relative and absolute likelihoods of radiative transitions from one state to another. Somewhat unexpectedly, however, even with all the information at hand (the two wave functions for the states involved in the transitions) we are unable to predict exactly when any particular transition will occur. Since the most we can know about the behavior of the electron is the probability of it being in a given place at a given time, all we are able to calculate is what the probability per unit time is for any particular transition to happen. Although this lack of knowledge may seem unsatisfactory, it is sufficient to predict quite accurately for a large collection of atoms in the same quantum state how many will de-excite in a given time interval. If the probability per unit time for a transition between states for an individual atom is some number, say p_0, then the number of atoms from a collection N that will undergo a transition in a time dt is

$$dN = - p_0 N \, dt$$
$$\frac{dN}{N} = - p_0 \, dt$$

This relation is familiar, and we know that it gives a time behavior

$$N(t) = N_0 \, e^{-p_0 t}$$

where N_0 is the number of atoms that had not undergone the transition at time $t = 0$. The decay is an exponential function of time and is characteristic of all quantum system transitions. The quantity $1/p_0$ is called the meanlife, τ. For atomic systems τ is typically 10^{-8} sec, although it depends quite strongly on the detailed properties of the initial and final state involved.

Although we have been thinking of the de-excitation of the atom—the emission of radiation—the basic description is equally true for the reverse process—absorption of radiation. Whether or not a photon will be absorbed by an atom (assuming of course it has the correct energy to correspond to the excitation of the atom to another "allowed" state) can only be described by a probability, calculated in the same manner as for the emission of the photon. Clearly, the more atoms the photon passes, the greater the chance that it will be absorbed. Hence for a photon passing through a large collection of atoms, the probability that it will be absorbed is given by the product of the absorption probability per atom times the number of atoms encountered. The probability of absorption per atom is basically the same quantity as the probability of emission of that same photon from an excited atom. For absorption it is usually described as an effective area* (as if the absorption were a collision) which we will label σ.

For a small thickness dx of material the probability of absorption is given by

$$\frac{\sigma \, N_A \rho}{A} \, dx$$

where N_A is Avagadro's number, ρ is the density of the material and A is its gram-atomic weight. This means that if there is an initial flux of photons I the number absorbed in that thickness will be given by

$$dI_{\text{photon}} = -\frac{I \, \sigma \, N_A \rho \, dx}{A}$$

so once again we find an *exponential* function describing the attenuation of the photon beam

$$I(x) = I_0 \, e^{-\sigma N_A \rho x / A}$$

Frequently the product $\sigma N_A \rho / A$ for a particular material is called the absorption coefficient, μ (its inverse, $1/\mu$, is called the attenuation length) so that the expression becomes

$$I = I_0 \, e^{-\mu x}$$

The probability of absorption for photons depends very much on what atoms are present, but generally the higher the photon energy, the less likely it is to be absorbed per atom, and the further it penetrates. The general variation of μ with material for a given photon energy is a decrease of attenuation length with increasing Z of the element (more electrons per atom). Of course the attenuation length is also very sensitive to the presence of voids or cracks in a material, and therefore is extremely useful in detecting them. This characteristic attenuation can also be used for determining the thickness and thickness variations of various materials, ranging from paper to metals. Typical X-ray "photographs" of a wide range of materials are shown in Fig. 4-9; the X-ray sources used span a hundredfold energy range.

* See Appendix 1

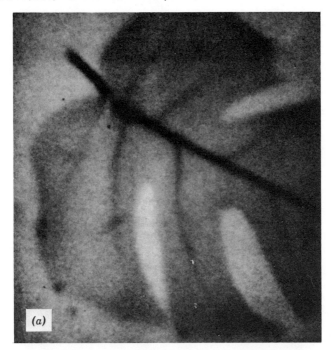

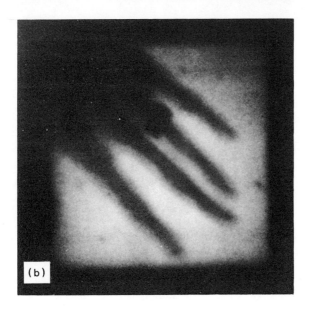

FIGURE 4-9 X-ray picture of: (*a*) Philodendron leaf, taken with 5.9 keV photons; (*b*) hand with ring, taken with 22 keV photons; (*c*) welds in boiler reheater tubes, taken with 660 keV photons. Leaf and hand courtesy of L. Kaufman, Department of Radiology, University of California School of Medicine, San Francisco; boiler reheater tubes courtesy Ontario Hydro.

In view of the wide range of material types and thicknesses that can be handled by high energy photons, or X rays, it is important to be able to produce X rays over a wide energy range. So far the only mechanism for X-ray production mentioned is by creation of vacancies in inner electron shells. This produces a set of discrete X-ray energies and is therefore of somewhat restricted usefulness. However, there is another mechanism for producing X rays which is used for many practical applications. This is called *bremsstrahlung* radiation.

If an energetic electron is driven at an atom it may strike one of the atomic electrons (as we discussed before), and cause the emission of characteristic X rays. However, the electron may not eject other electrons, but instead just be deflected in its path by the atomic nucleus. In undergoing the deflection, which of course is an acceleration, the electron may emit radiation, as we discussed in the classical picture of the atom. Now, since the electron trajectory in this case does not form a closed path around the nucleus, the electron wave properties that were so important in the role of generating a stable orbit no longer necessitate a standing wave condition, so the electron can follow any classically predicted trajectory. Consequently there can be a completely continuous range of energy radiated off by the electron. The general radiation spectrum due to a beam of energetic electrons striking a collection of atoms—a vacuum tube anode—will be a *continuum*, called bremsstrahlung radiation. (Of course there will be additional X rays due to excitation of K, L, etc., characteristic X rays of the anode material.) The maximum energy of the X-ray continuum corresponds to the maximum energy the electrons can lose, which is equal to their initial kinetic energy. By varying the initial kinetic energy of a beam of electrons that strike an X-ray production target, a wide range of X-ray energies can be produced. An arrangement for producing X rays is shown in Fig. 4-10, together with the resulting radiation spectrum.

An additional source of radiation is frequently used to obtain very high photon energies, typically from several tens to thousands of keV. These high energy X rays, called γ rays, are due to the radioactive decay of a range of elements (which will be discussed in sec. 9-5). A practical advantage of this kind of source is that it requires no external power source.

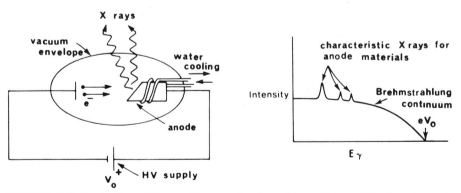

FIGURE 4-10 Production of X rays. Electrons decelerated by collisions with atoms in anode produce an X-ray continuum up to a maximum energy equal to the electrons' initial kinetic energy eV_0. Atomic excitations also produce discrete X rays corresponding to the atom level structure.

4.4 FINE STRUCTURE; ELECTRON SPIN In principle, having solved the Schrödinger equation for hydrogen, we should be able to predict all the energy levels of the hydrogen atom and therefore all possible transitions in the hydrogen emission spectrum. However, this is true only if we do not look with very high resolution; with high resolution many of the transitions are found to be closely spaced doublets. The existence of these extra transitions is inexplicable in terms of our description based on the quantum numbers n, ℓ and m_ℓ. In fact the observation of the spectral doublets led to the proposal of an additional electron property; specifically that the electron has internal (intrinsic) angular momentum which is quantized.

It is not immediately obvious why the presence of unexpected doublets should imply the existence of internal angular momentum nor even that the electron ought to have internal angular momentum at all. However it is not difficult to realize that the Schrödinger equation might be inadequate in some way, particularly when we remember that it makes use of nonrelativistic assumptions, and when we realize that electrons, particularly when tightly bound, may not satisfy such an assumption. P. A. M. Dirac was able to develop relativistic wave equations for the electron, and found that the electron wave equation has additional degrees of freedom which show an identical behavior to that which an object with angular momentum would have.

Proof of the existence of the electron's internal angular momentum, called *spin*, was obtained by direct observation of different trajectories caused by the possible spin orientations in an external magnetic field (Stern-Gerlach experiment*).

To understand how the Stern-Gerlach experiment gives proof of the electron spin and to see how the existence of spin gives rise to doublets in the observed optical transitions we must consider the properties of a magnetic dipole in a magnetic field. We will look specifically at the origin of the doublets; the so-called spin-orbit interaction.

The existence of spin, or internal angular momentum, implies the existence of charge in motion "within" the electron. By the same arguments made for the Bohr magneton we know that the electron must therefore have an intrinsic magnetic dipole moment, due to the existence of internal angular momentum. If the electron also has orbital angular momentum (i.e., it has a nonzero value of ℓ) that orbital motion about the positively-charged nucleus will give rise to a magnetic field, in which the electron exists. (It may be easier to think of this from the point of view of the electron; if $\ell \neq 0$, the nucleus, with charge $+z$ is orbiting around the electron and *it* creates the magnetic field that the electron feels.) A magnetic dipole in a magnetic field has an energy

$$E_m = - \boldsymbol{\mu} \cdot B$$

where μ is the electron's intrinsic magnetic moment. This term will modify the total energy of the electron in the atom. In principle one ought to include this term in the Schrödinger equation and solve it again. However, since the magnetic energies are

* See Further Reading, Tipler, sec. 7-4.

several orders of magnitude smaller than those due to the Coulomb term it is reasonable to treat this effect as a small perturbation on the system and simply correct the energy of the levels* by the term E_m instead of solving the complete Schrödinger equation. By analogy to the Bohr magneton the intrinsic angular momentum should be proportional to the internal angular momentum

$$\mu_{\text{int}} = -g\,\frac{e\,\mathbf{S}}{2m_e}$$

where g is a proportionality constant (called the gyromagnetic ratio or g factor) and $\mathbf{S}$ is the internal or intrinsic electron angular momentum, which has associated with it a "spin" quantum number s, according to the relation

$$\mathbf{S} = \sqrt{s(s+1)}\,\hbar$$

Therefore the magnetic energy will be

$$E_m = g\,\mu_B\,m_s\,B_z$$

where m_s is the projection of the internal angular momentum on the z axis and μ_B is the Bohr magneton; in this case the z axis is the direction of the magnetic field. The internal angular momentum quantum number is s, and is analogous to ℓ. The number of possible orientations is given by the possible values of m_s and equals $2s + 1$, which corresponds to the number of distinct energy levels the internal magnetic field will "create." The observed effect was a doubling, and therefore $s = 1/2$. The electron is said to have spin one half (this is also the value predicted by Dirac).

The value of g can be obtained from the magnitude of the energy splitting, or in principle from the Stern-Gerlach experiment where the magnetic field is directly measurable. The result is that for the electron $g = 2$, so that the electron magnetic moment is

$$|\mu_e| = \frac{e\hbar}{m_e}\,\sqrt{\tfrac{1}{2}\,(\tfrac{1}{2}+1)}$$

The fact that $g = 2$ and not 1 (as it is for orbital angular momentum) can only be explained as a relativistic effect.

We must now add two more quantum numbers to our quantum description of the state of an electron in hydrogen; s and m_s. This complicates matters a bit whenever the spin-orbit interaction is present since the mutual interaction is equivalent to a torque that changes the orientation of both L and S. Therefore m_ℓ and m_s are no longer good quantum numbers because they cease to be constants of motion. However, there can be no torque on the sum of the orbital and intrinsic angular momentum, which we will call the *total* angular momentum $\mathbf{J}$, which has a corresponding quantum number j and z projection quantum number m

* This assumption becomes invalid in cases where magnetic effects are large.

$$\mathbf{J} = \mathbf{L} + \mathbf{S}, \qquad |\mathbf{J}| = \sqrt{j\,(j+1)}\,\hbar$$

with $\qquad j = \ell \pm \frac{1}{2}, \qquad m = m_\ell + m_s$

Although it is now strictly true* that j and m are the only "good" angular momentum quantum numbers, we know that the spin-orbit interaction introduces only a small perturbation of the system, so we can expect that L and S still represent valid concepts. Therefore we will continue to use ℓ, s, m_ℓ, and m_s where they serve our purpose, provided that the particular physical situation still warrants their use.

Because of the electron intrinsic spin the nomenclature of the levels now becomes somewhat more complex. For instance the $n = 2$, $\ell = 1$, $j = \frac{1}{2}$ level is normally described by the symbol

$$n = 2 \longrightarrow 2\,P_{1/2} \overset{\ell = 1}{\underset{j = \frac{1}{2}}{\longleftarrow}}$$

The ground state of hydrogen ($n = 1$, $\ell = 0$, $j = \frac{1}{2}$) then becomes $1s_{1/2}$ in its total quantum description. This scheme is illustrated for hydrogen in Fig. 4-11.

4.5 APPLICATION OF ATOMIC INTERACTIONS; THE ZEEMAN EFFECT

It is a reasonable statement that any effect, or interaction, can be exploited to develop a device. We saw that the internally-produced magnetic field interacting with the electron spin caused a change in the energy of the electron in hydrogen. In fact *any* applied field, magnetic or electric, will produce a change in the level structure of any

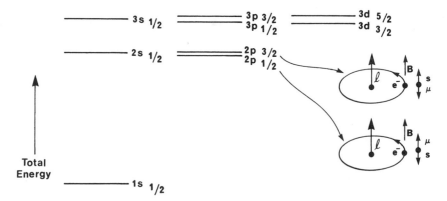

FIGURE 4-11 Level structure of hydrogen including spin-orbit splitting (not to scale). It is conventional to use lower case letters for angular momenta of individual electrons and upper case letters for the angular momenta of whole atoms. In the case of hydrogen the two are the same.

* For a detailed discussion, see Further Reading, Leighton, sec. 5-6.

atom, and if this change can be observed and quantitatively measured, the effect can be "turned around" and used as a device to detect and measure the inducing field. The Zeeman effect provides a good example of this.

The Zeeman effect is the change of the energy levels of an atom produced by an externally applied magnetic field (analogous to the spin-orbit interaction). The change of energy will be, as before,

$$\Delta E = E_m = -\boldsymbol{\mu} \cdot \mathbf{B}$$

where $\mathbf{B}$ is an externally applied field. This means for instance that the ground state energy of hydrogen which has a quantum number $j = \frac{1}{2}$ will be changed, depending on the orientation of this spin in an applied magnetic field

$$E = E_0 + \Delta E$$

$$= E_0 - g\,\frac{e\hbar}{2m_e}\,m_j\,B$$

Now the possible projections of the magnetic moment (which is antiparallel to the spin j) for a $j = \frac{1}{2}$ state on *any* axis, hence the axis in the direction of $\mathbf{B}$ can only be $+\frac{1}{2}$ or $-\frac{1}{2}$. Therefore in the presence of a magnetic field the ground state of hydrogen will be split into two separate states, as shown in Fig. 4-12.

Any hydrogen atom that happens to have its spin *up* (parallel to the direction of the magnetic field) when the magnetic field is turned on will become slightly less tightly bound, by an amount

$$\Delta E = \frac{e\hbar B}{2\,m_e}$$

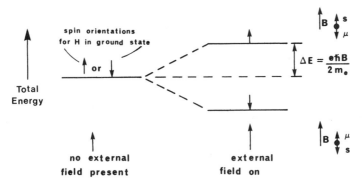

FIGURE 4-12 Effect of external magnetic field on ground state energy of a collection of hydrogen atoms.

Any atom that has spin down will have its energy shifted down by an equal amount. Atoms that find themselves in the upper of the two possible states will radiate off their excess energy and settle down into the lower state. Then it should be possible to *induce* a transition from the lower to the upper state by "shining" radiation on the hydrogen at a frequency corresponding to the energy difference of the two states.

$$h\nu = E_{upper} - E_{lower}$$

$$= E_0 + \frac{e\hbar B}{2 m_e} - \left(E_0 - \frac{e\hbar B}{2 m_e} \right)$$

$$= \frac{e\hbar B}{m_e}$$

$$\nu = \frac{eB}{2\pi m_e}$$

Absorption of radiation at this frequency will cause those atoms in the lower level to be excited to the higher one. To what frequency does this correspond for a magnetic field of the order of, say, the earth's ($\sim 10^{-4}$ T)?

$$\nu = \frac{1.6 \times 10^{-19} \times 10^{-4}}{2\pi \times 10^{-30}} \cong \frac{2}{6} \times 10^{30-23} \cong 3 \ MH_z$$

If we pump radio frequency (*RF*) energy into atomic hydrogen (or any quantum system with $j = \frac{1}{2}$) the energy will be absorbed at the frequency that corresponds exactly to the energy difference of the two levels. The net result will be an increased loss of *RF* power to the hydrogen at that frequency. This fact can be used to measure the magnetic field very accurately. The range of magnetic fields measurable with such a device is limited largely by the range of oscillator frequencies available. This is something like 10 kHz → 100 MHz, a range of more than six orders of magnitude.

It should also be possible to observe the de-excitation radiation. In fact if there is any thermal energy available it should be possible to thermally "pump" the atoms to the magnetic excited state and measure the frequency of the emitted radiation without having to supply additional energy. Such a mechanism might be useful in measuring the magnetic field in a very hot gas, or plasma, in a magnetic field. In such systems there exists the possibility of populating excited states ($n > 1$), and consequently more complex spectra will be produced. The levels excited split into $2j + 1$ different levels in the magnetic field* (Fig. 4-13), corresponding to the possible orientations of the magnetic moment with respect to the magnetic field direction. In the case of a general spin state, the magnitude of the splitting is

---

* The following treatment assumes that the magnetic field only produces a small perturbation in the system.

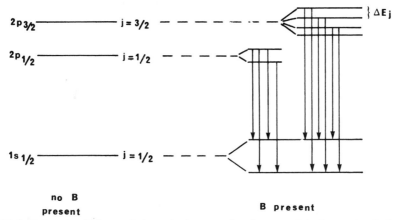

FIGURE 4-13 Level splitting and de-excitation energies for several hydrogen levels in an external magnetic field. The magnitude of the level splitting ΔE_j depends on the total spin j.

$$\Delta E_j = E_{m_j} - E_{m_{j-1}}$$

$$= \frac{ge\hbar}{2m} B[m_j - (m_j - 1)]$$

$$= \frac{ge\hbar B}{2m}$$

for any two adjacent states. However the numerical value of g now depends on the value of j, ℓ, and s, and has to be generalized to

$$g = 1 + \frac{j(j+1) + s(s+1) - \ell(\ell+1)}{2j(j+1)}$$

because the total magnetic moment consists of the vector sum of the moment due to orbital angular momentum (for which $g=1$) and the moment due to intrinsic spin (for which $g=2$). This general formula reduces to $g=2$ for $j=\frac{1}{2}$, $\ell=0$, as we expect.

There are, of course, more bizarre ways in which the Zeeman effect might be used. For example it ought to be possible to measure changes in the number of atoms that can absorb the exciting radiation, by monitoring the absorbed power. Such a technique, outlined in Fig. 4-14, has been shown to be feasible for measuring the volume of bubbles in boiling water, and also giving an accurate indication of this fraction as a function of time.

4.6 MULTIPLE ELECTRON ATOMS; THE PAULI PRINCIPLE So far we have only dealt properly with single-electron atoms. We do not really know how to put more than one electron into an atom. We do not know, for instance, any reason

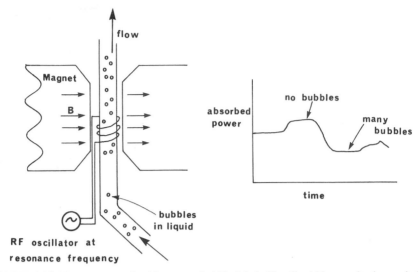

FIGURE 4-14 Measurement of void content (bubbles) in boiling liquid by monitoring variation of *RF* power absorbed by liquid in tube. A reduction of absorbed power indicates an increased void fraction in the liquid. After G.F. Lynch and S.L. Segel, *Int. J. Heat Mass Transfer 20* (1977), 7.

why the Moseley plots of X rays should find that there are only 2 electrons in the K shell and 8 electrons in the L shell.

What would we expect to happen if we put two electrons into the same atom? They will presumably both be attracted to the nucleus and end up in the lowest possible quantum state available to them. Does this mean then that they will both end up in the same state $(n, \ell, m_\ell, s, m_s)$? What would it mean if they did? If the two electrons were to end up in exactly the same quantum state there would be absolutely no way that we could distinguish between the motions of those two electrons. We could not say, for instance, that the two electrons were in the same place at the same time, which we know classically must not happen. However, these electrons are not "classical" so we must consider the situation more carefully.

There is a basic difference between the classical and quantum-mechanical view of matter. Classically it is always possible (in principle at least) to trace the path of an individual particle through a collection of otherwise identical particles. This is because we can always know both the position and momentum of a classical particle infinitely accurately so that the trajectory of any particle can be determined sufficiently accurately to avoid mistaking one particle for another—the particles are *distinguishable*. However, for *real* (quantum mechanical) particles, the uncertainty principle forces us to abandon the expectations of being able to distinguish between particles since we know we cannot follow a trajectory accurately enough to be sure we have not "lost track." A quantum description of such a system must be written in such a way that we are sure that the uncertainty principle has not been violated by being able to distinguish individual particles.

This means, for example, that if we tried to characterize a system of two identical particles with an overall wave function Ψ_a by

$$\Psi_a = \Psi_1(r_1)\Psi_2(r_2)$$

where $\Psi_1(r_1)$ gives the probability amplitude for particle 1 being at position r_1 and $\Psi_2(r_2)$ gives the probability amplitude for particle 2 being at position r_2, we would be violating the uncertainty principle, since we could then distinguish between that situation and the one given by a second overall wave function Ψ_b,

$$\Psi_b = \Psi_2(r_1)\ \Psi_1(r_2)$$

which is identical to Ψ_a, except that particles 1 and 2 have been interchanged. The uncertainty principle demands that the particles cannot be identified unambiguously; they must be indistinguishable. This can be satisfied by using a linear combination of Ψ_a and Ψ_b of the form

$$\Psi = \frac{1}{N}[a\Psi_1(r_1)\Psi_2(r_2) \pm b\ \Psi_2(r_1)\ \Psi_1(r_2)]$$

where N is a normalization constant which ensures that the total probability of finding a particle is unchanged in the combined wave function, and a and b are constants. It is observed, and can be shown to be a necessary consequence of relativistic quantum mechanics, that a and $b = 1$, and that the *plus* sign in the formula must be used when the intrinsic spin of the particles being described is *integral* (such as a photon); the *minus* sign must be used when the intrinsic spin of the particles is *half-integral* (such as an electron). The general class of integral spin particles are called *bosons* and half integral particles are called *fermions*.

If the wave function for integral spin particles (bosons)

$$\Psi = \frac{1}{N}[\Psi_1(r_1)\Psi_2(r_2) + \Psi_2(r_1)\ \Psi_1(r_2)]$$

is examined it can be seen that interchanging the labels 1 and 2, which is equivalent to exchanging particles 1 and 2, does not alter the overall wave function Ψ; it is *symmetric* under exchange of particles and accordingly is called a symmetric wave function. In contrast, the wave function for half integral spin particles (fermions)

$$\Psi = \frac{1}{N}[\Psi_1(r_1)\ \Psi_2(r_2) - \Psi_2)r_1)\ \Psi_1(r_2)]$$

changes overall sign on interchanging labels and is described as an *antisymmetric* wave function.

We are now in a position to work out what the answer to our original question must be. If we allow the two electrons that we asked about to be in the same atom and in the same $(n, \ell, m_\ell, s, m_s)$ state, then Ψ_1 and Ψ_2 would be identical functions (which we will simply label Ψ). In that case the total wave function would have to be

$$\Psi = \frac{1}{N}[\Psi(r_1)\,\Psi(r_2) - \Psi(r_1)\,\Psi(r_2)]$$

which is identically zero everywhere. Therefore they *cannot* exist in exactly the same state, as our intuition would have led us to guess. However, the situation is more subtle than we might be inclined to believe from this success, since it is in fact possible for those two electrons to be in states that are identical in all respects except, for example, their intrinsic spin projection. We could not have guessed that one arrangement would be possible and the other not. Classical intuition is good as a general guide, but not for ultimate details.

The general statement that no two identical half-integral particles in the same quantum system can have all their quantum numbers identical was first surmised by W. Pauli and consequently is called the *Pauli Exclusion Principle*. Stated in positive terms, this principle requires that there be at least one quantum number that is different for every electron (fermion) in a quantum system.

This general principle can be applied to the question of what quantum states electrons can populate in any particular atom, that is how many electrons we can put into any particular n, ℓ quantum set of levels. Clearly the number of electrons that can exist together in any n, ℓ quantum level is the total number of different ℓ "orientations" and s "orientations" that are possible. This number is $(2\ell+1)(2s+1) = 2(2\ell+1)$ for a particular ℓ value. Therefore for the $n = 1$ level we can satisfy the Pauli principle with no more than 2 electrons, that is,

	$n,$	$\ell,$	$m_\ell,$	$s,$	m_s	
	1,	0,	0,	½	$+$ ½	
and						2
	1,	0,	0,	½	$-$ ½	

For $n = 2$ this becomes

$n = 2 \qquad \ell = 0$

	2,	0,	0,	½	$+$ ½	
						2
	2,	0,	0,	½	$-$ ½	

$n = 2,$ $\quad \ell = 1$

$$
\left.
\begin{array}{ccccc}
2, & 1, & 0, & \frac{1}{2} & + \frac{1}{2} \\
2, & 1, & 0, & \frac{1}{2} & - \frac{1}{2} \\
2, & 1, & 1, & \frac{1}{2} & + \frac{1}{2} \\
2, & 1, & 1, & \frac{1}{2} & - \frac{1}{2} \\
2, & 1, & -1, & \frac{1}{2} & + \frac{1}{2} \\
2, & 1, & -1, & \frac{1}{2} & - \frac{1}{2}
\end{array}
\right\} \quad 6
$$

Notice that this says there can be 2 electrons in the $n=1$, or K shell, and $2+6 = 8$ electrons in the $n = 2$ or L shell. These are exactly the numbers determined from the Moseley X-ray plots. The full quantum description of the atom, including the electron intrinsic spin, reproduces the basic shell picture used in the shielded Bohr model calculations. However the "shells" are hardly shells any more, since the electrons are considerably spread out in space, as the radial distributions show us. In addition, for quantum levels where $\ell \neq 0$, which we would expect not to show spherical symmetry, we find that when *all* the magnetic substates (that is, values of m_ℓ) are filled the charge distribution of the corresponding electrons is once again spherically symmetric.

This latter fact leads to a very important consequence of the Pauli principle for multielectron atoms. Since a spherically symmetric electron distribution in an atom has the sole effect of canceling an equivalent amount of positive charge in the nucleus, such a distribution is ineffective at attracting or repelling external charges. Therefore the atom can be "active" only in terms of any interaction involving the Coulomb force (e.g., chemically) if there is an unfilled shell. Consequently inner shells (ones with lower values of n than the maximum value occupied) can only contribute if one or more electrons are removed from that inner shell, either through ionization or excitation to higher shells. Under most circumstances this means that only electrons in unfilled shells, or at the very most the outermost filled shell, are likely to take part in Coulomb interactions or be excited to higher levels since these take by far the least energy to excite. Let us take sodium as an example. It has all levels filled to the $n = 2$ shell and one electron in the $3s$ level when in its ground state. We will compare the energy required to excite a $3s$, $2s$, or $1s$ electron to a higher-energy unoccupied level; for example, the $4s$ level

$$
\Delta E_{3s \to 4s} \cong -Z_{\text{eff}}^2 \, E_0 \left(\frac{1}{4^2} - \frac{1}{3^2} \right).
$$

$$
= - E_0 \left(\frac{1}{16} - \frac{1}{9} \right) = 0.05 \, E_0 = 0.7 \, eV
$$

since the effective nuclear charge that the $3s$ electron "sees" is approximately 1. To excite an electron from the $2s$ level, however, we must remove an electron which sees the full nucleus $Z = +11$ shielded essentially *only* by the two $1s$ electrons

$$
\Delta E_{2s \to 4s} \cong + \frac{(11-2)^2 \, E_0}{2^2} - \frac{(1)^2 \, E_0}{4^2} \cong 20 \, E_0 = 275 \, eV
$$

Already the energy required has increased by two orders of magnitude. If the $1s$ electron is to be excited, it will require an energy of the order

$$\Delta E_{1s \to 4s} \cong (11)^2\, E_0 - \frac{1^2\, E_0}{4} = 120\, E_0 = 1.64\ keV$$

which is even greater. Clearly all chemical and optical properties are most likely to be determined *only* by the outermost, least tightly bound electrons. This statement is no more than the statement made by chemists that the chemistry of an atom is determined by its outermost, "valence" electrons. In fact, as can be seen in Fig. 4-15, the whole periodic table of elements can be thought of correctly as a listing of elements in order of increasing nuclear charge, grouped according to the orbital angular momentum of their outermost electrons.

For high-Z materials the filling order of the shells some irregularities (see V − Cr − Mn − Fe − Co − Ni − Cu in Table 4-1). However, these are explained fully by a detailed quantum mechanical treatment.

The concept of active outer electrons is also valid for interpreting spectroscopic information. As can be seen from Fig. 4-16, the spectrum of sodium is very similar to the spectrum of hydrogen. The valence electron "sees" effectively only a single net positive charge of the nucleus. The detailed treatment of core shielding is more complex than the simple arguments used in sec. 3-6 due to the different radial wave functions of the various states, but the gross features remain correct.

Now, in principle at least, the calculation of the properties of an isolated atom of any element is completely soluble. The combination of the Schrödinger equation and the Pauli principle enable us to solve the problem, although the Bohr model still provides a valuable qualitative picture as well as a way of making quantitative energy estimates.

FURTHER READING   A. Beiser, *Concepts of Modern Physics*, 2nd ed., McGraw-Hill, 1973, Ch. 7.

R. B. Leighton, *Principles of Modern Physics*, McGraw-Hill, 1959, Ch. 5.

G. F. Lynch and S. L. Segel, Direct Measurement of the Void Fraction of a Two-Phase Fluid by Nuclear Magnetic Resonance, *International Journal of Heat Mass Transfer*, No. 20 (1977), p. 7.

J. Norwood, *Twentieth Century Physics*, Prentice-Hall, 1976, Ch. 9.

P. A. Tipler, *Modern Physics*, Worth, 1978, Ch. 7.

A. L. Reiman, *Physics, Vol. III: Modern Physics*, Harper-Row, 1978, Ch. 8.

M. R. Wehr, J. A. Richards, and T. W. Adair, *Physics of the Atom*, 3rd ed. Addison–Wesley, 1978, Ch. 6, X-Rays.

PROBLEMS   1. List the quantum numbers $(n,\ \ell,\ j)$ of what you would expect the first eight excited states of hydrogen to be from all the known properties of the hydrogen atom.

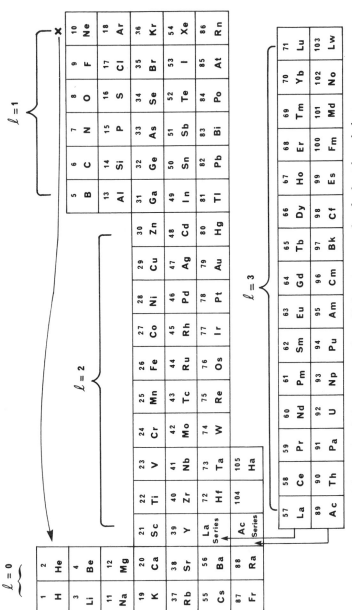

FIGURE 4-15 Table of elements, together with l quantum number for least bound electrons.

Table 4-1   ELECTRON ORBITAL FILLING ORDER

	K	L		M			N				O				P			Q
	1s	2s	2p	3s	3p	3d	4s	4p	4d	4f	5s	5p	5d	5f	6s	6p	6d	7s
1 H	1																	
2 He	2																	
3 Li	2	1																
4 Be	2	2																
5 B	:	2	1															
6 C		2	2															
7 N		:	3															
8 O			4															
9 F			5															
10 Ne			6															
11 Na			6	1														
12 Mg			6	2														
13 Al			:	2	1													
14 Si				2	2													
15 P				:	3													
16 S					4													
17 Cl					5													
18 A					6													
19 K					6		1											
20 Ca					6		2											
21 Sc					:	1	2											
22 Ti						2	2											
23 V						3	2											
24 Cr						5	1											
25 Mn						5	2											
26 Fe						6	2											
27 Co						7	2											
28 Ni						8	2											
29 Cu						10	1											
30 Zn						10	2											
31 Ga						10	2	1										
32 Ge						:	2	2										
33 As							:	3										
34 Se								4										
35 Br								5										
36 Kr								6										
37 Rb								6			1							
38 Sr								6			2							
39 Y								6	1		2							
40 Zr								:	2		2							
41 Nb									4		1							
42 Mo									5		1							

Table 4-1 (Continued)

	K	L		M			N				O				P			Q
	1s	2s	2p	3s	3p	3d	4s	4p	4d	4f	5s	5p	5d	5f	6s	6p	6d	7s
43 Tc									5		2							
44 Ru									7		1							
45 Rh									8		1							
46 Pd									10									
47 Ag									10		1							
48 Cd									10		2							
49 In									10		2	1						
50 Sn									10		2	2						
51 Sb									10		2	3						
52 Te									10		2	4						
53 I									10		2	5						
54 Xe									10		2	6						
55 Cs									10		2	6			1			
56 Ba									10		2	6			2			
57 La									10		2	6	1		2			
58 Ce									10	2	2	6			2			
59 Pr									10	3	2	6			2			
60 Nd									⋮	4	2	6			2			
61 Pm										5	2	6			2			
62 Sm										6	2	6			2			
63 Eu										7	2	6			2			
64 Gd										7	2	6	1		2			
65 Tb										9	2	6			2			
66 Dy										10	2	6			2			
67 Ho										11	2	6			2			
68 Er										12	2	6			2			
69 Tm										13	2	6			2			
70 Yb										14	2	6			2			
71 Lu										14	2	6	1		2			
72 Hf										14	2	6	2		2			
73 Ta									⋮	⋮		6	3		2			
74 W												⋮	4		2			
75 Re													5		2			
76 Os													6		2			
77 Ir													7		2			
78 Pt													9		1			
79 Au													10		1			
80 Hg													10		2			
81 Tl													10		2	1		
82 Pb													10		2	2		
83 Bi													10		2	3		
84 Po													10		2	4		

2. If atoms could contain electrons with principal quantum numbers up through $n = 6$, how many different elements would there be?

3. Prove that the probability of finding an electron between the radial distance r and $r + dr$ is $r^2 R_{n\ell}$.

4. Estimate the size of He, Ne, Na, and singly-ionized Li atoms using the Bohr model. How do these estimates compare with what is observed?

5. The levels in a hypothetical single-electron atom are given by;

n	1	2	3	4	5	∞
$E_n(eV)$	-30.60	-7.65	-3.40	-1.50	-0.98	0

Draw the level diagram and find (a) the energy required to ionize the atom, (b) the energy required to excite the atom from the ground state to the n=3 level, (c) the frequency of the photon emitted when the n=3 level deexcites to the ground state, (d) the highest level for which the Bohr model works, and (e) the nuclear charge.

6. Predict what the relative ordering of the 2s and 2p levels should be in Li using the radial electron probability distribution in Fig. 4-5 and the fact that in H they are at the same energy. Use the Bohr picture of multielectron atoms to treat the inner shell electrons. Does this prediction agree with what is observed?

7. Both hydrogen and sodium can be thought of as single-electron atoms, and therefore should have similar level structures. However, it is found that although there are similarities, the ionization energy (the energy required to remove an electron) for sodium is much less than for hydrogen, and in sodium levels with the same n but different ℓ quantum numbers have different energies whereas in hydrogen they have the *same* energy. Explain why these differences occur.

8. Describe some physical circumstances where you would expect an electron wave function to be composed of a linear combination of $+m$ and $-m$ solutions in preference to a unique value of m.

9. What fraction of a collection of initially excited atoms are still excited after a time 5τ?

10. The uncertainty principle requires that $\Delta E \Delta t > \hbar/2$. A collection of atoms in an excited state is characterized by a lifetime τ, which characterizes the range of times that an individual atom exists before de-exciting. What is the corresponding range of energies at which the excited atoms may exist if $\tau = 10^{-8}s$? What is it if $\tau = 10^{-16}s$?

11. Estimate the energy of the $3s \rightarrow 2s$ and K_α transition in lithium using the Bohr model. How accurate are the predictions?

12. Calculate the expected continuous X-ray spectrum end point and the highest-energy intense characteristic X-ray energy for a beam of 50 keV electrons striking a copper target.

13. You have been given the task of setting up a system to examine thick metal tubes for faulty welds and small cracks. Assuming you could produce the same energy

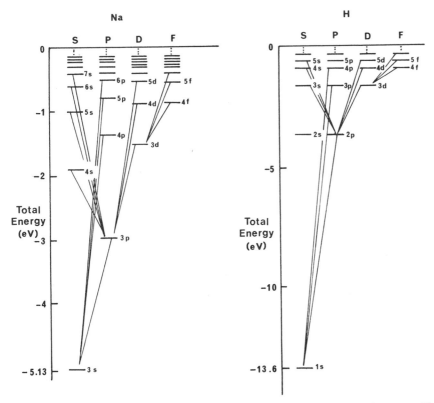

FIGURE 4-16 Comparative level scheme of Na (valence 1) and H ($Z=1$) atoms. Note the difference in energy scales.

X rays with an X-ray tube and a radioactive source, which would you choose in order to obtain the highest possible resolution pictures? Why?

14. At extremely high photon energies (in the MeV region) the characteristic cross section for photons interacting with atoms should be of the order of 10^{-18}cm^2. What thickness of copper would you expect to have to use to cause 50 percent of a beam of such photons to have interacted?

15. The bombardment of materials with beams of very energetic particles is known to cause electron rearrangements which lead to changes in the chemistry of elements. If an atom of chlorine is struck such that a $3s$ electron is promoted to the valence level, what chemical element would this modified atom resemble? Using the Bohr model estimate the total energy of this "new" element compared to the one it resembles.

16. A hydrogen atom is placed in an external magnetic field of 0.5 Tesla. What are the energies (in eV) of all distinct $n = 1$ and $n = 2$ levels?

17. The magnetic field B inside a hydrogen plasma confinement system is to be measured under running conditions and, if possible, by a remote sensing technique. When the magnetic field is turned on and the hydrogen gas is heated, a strong emission at 2.8×10^{10} Hz is observed.

(a) What is the value of B?

(b) As the gas is heated even more, a second frequency is observed at ⅓ the previous value. What is this due to?

(c) Can you predict whether or not this technique would work at room temperature?

18. The proton and electron are both charged, spin ½ particles. Predict the magnetic moment of the proton using the Bohr model.

19. Impurities in a plasma introduced by ion bombardment of the vacuum chamber walls are known to cause significant problems when attempting to heat a plasma. You have been asked to investigate the possibility of detecting the presence of lithium atoms in the vacuum system residual gas by a magnetic resonance technique. A 1 cm^3 volume, immersed in a 0.3 Tesla field at the edge of the confinement system, is available for testing. At what frequency will the magnetic resonance signal occur? If the gas is at 300 K and 10^{-9} of an atmosphere of lithium atoms are present how much energy could be absorbed by the sample (assume no atoms re-emit the absorbed energy)?

FIVE

THE QUANTUM MOLECULE

Having solved the problem of the quantum atom, we must now recognize that almost all cases of practical interest will involve not isolated atoms but collections of atoms—either as molecules or as solids. We therefore have to look at our quantum atom and see how these other forms of matter can arise. We will start by trying to make a molecule.

5.1 FORMATION OF MOLECULES The simplest system to consider is H_2^+; that is, a molecular hydrogen ion, or two protons and one electron (Fig. 5-1).

If we wanted to seriously calculate what happens in this system, we would set up the Schrödinger equation for a hydrogen atom and then add to it a repulsive potential due to the V_{++} (proton-proton interaction), an attractive potential due to the V_{+-} (electron-proton interaction) and kinetic energy terms for the two protons. This would be a difficult problem to solve. However, qualitatively we can see easily what should happen. Let us look at the ground state wave function of the electron as a function of the separation of the two protons. This is illustrated in Fig. 5-2.

For large separations* of the protons the ground state electron wave function is a linear combination of the wave functions for the electron being bound to each of the atoms. For large separations these are approximately the $1s$ solution for an isolated hydrogen atom. As the two protons are brought closer together so that the individual wave functions start to overlap strongly, we find that there is an enhanced probability of the electron being found between the two protons, and so we expect the strong attraction of the two protons to the electron to overcome their mutual repulsion and form a molecule. We can also do a rough calculation to confirm this. When the

* Compared to the atom size.

105

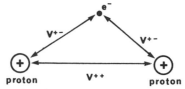

FIGURE 5-1 Coulomb interactions in a hydrogen molecular ion.

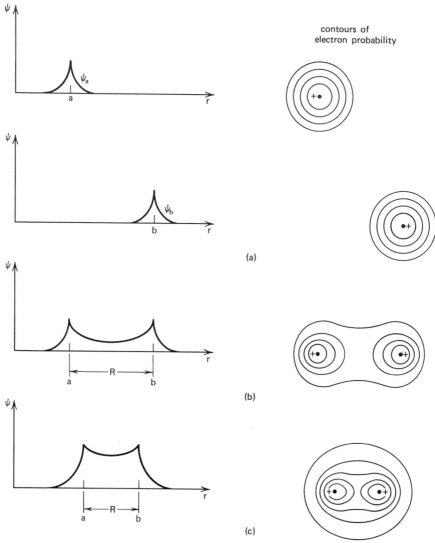

FIGURE 5-2 Electron wave function in the presence of two protons (symmetric combination) (*a*) at large separation, (*b*) at intermediate separation, and (*c*) at a separation comparable to the wave function spread.

separation R between the two protons is infinite the energy of the electron must be -13.6 eV, that is, it must be bound to a proton as a hydrogen atom. If we let R go to 0 then the electron is in a helium atom and provides an energy

$$E_{el} = -Z^2 E_0 = 4 \times 13.6 = -54 \ eV$$

The total energy of the system will be the sum of the electron attractions and the proton-proton repulsion, which is $V_{++} = k \ e^2/R$. The variation of E_{el} with R is more difficult to calculate, but will change quite rapidly from ~ -13.6 eV to -54 eV in the region where the electron wave function is shared strongly between the two protons.

Clearly for very small separations the sum of the two contributions is dominated by the repulsive term and will be positive, but for slightly greater separations the attraction will cause a net *decrease* of total energy, and provide a *binding energy* to make a molecule. As shown in Fig. 5-3, there is a net binding of approximately 2.6 eV for the H_2^+ molecule, and an equilibrium separation of about 1×10^{-10}m. The case for neutral hydrogen can be dealt with in the same way, except that we have the complication of the second electron which will provide both more attraction for the protons

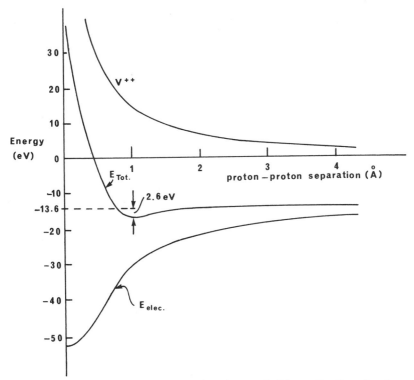

FIGURE 5-3 Electronic, proton repulsion and total energy in H_2^+ as a function of proton-proton separation for symmetric wave function.

(about 2.6 eV) and a small amount of repulsion between the two electrons, (about 0.8 eV). By necessity their spins are antialigned so they can be in the same spatial state (i.e., both have the same symmetric spatial wave function).

The above is not the only possible H_2^+ state produced by a combination of the two $1s$ wave functions. It is also possible to start with a linear combination of the two wave functions where their relative signs are opposite and obtain an additional solution of the Schrödinger equation. This combination of the wavefunctions, shown in Fig. 5-4, is labeled antisymmetric, whereas the ones in Fig. 5-3 are labeled symmetric (see sec. 4-6).

When the two protons are brought together the antisymmetric combination of wave functions begins to cancel at the midpoint between the protons, so that an electron in that state does not provide the additional attraction necessary for binding the two atoms together. When brought completely together this would look like helium in an excited state, since the combined wave function is no longer a $1s$ state. The net result, shown in Fig. 5-5, is that no bound state will be formed, and a net repulsive "force" will exist for all atom-atom separations.

5.2 ADDITIONAL DEGREES OF FREEDOM; ROTATION AND VIBRATION
Up to now all the motion, or all the possible forms of energy of the system has been associated with the electrons alone. Now, however, by binding two rather massive atoms together we have introduced the possibility of other kinds of motion being present (other means of storing energy). If we imagine a molecule as a semirigid dumbbell, it is quite clear that in principle we ought to be able to store additional energy by allowing the atoms to rotate around each other; also, by displacing the atoms in a molecule from their equilibrium separation, electrostatic potential energy can be stored and subsequently released as kinetic energy, that is, vibrational energy.

These are two additional ways of storing energy—additional degrees of freedom for the system. In principle they are represented in the Schrödinger equation through the proton kinetic energy terms. Again, rather than attempt to make a "proper" treatment of this problem, we shall only attempt to get a rough idea of the energy associated with these two additional types of motion compared to the internal, or electronic, energies.

To estimate the energy of rotational motion we will use the same approach that was so successful in first treating the atom. We will start as we did with the Bohr model by first making a classical picture of the motion. It is quite clear that the rotational motion of an atom is of the same nature as an "orbiting" electron, which means that a wave description for the atom's motion—the standing wave condition—will once again become important. The standing wave condition can be incorporated into the energy description by quantizing angular momentum since we found in the development of the Bohr atom that the two conditions are equivalent.

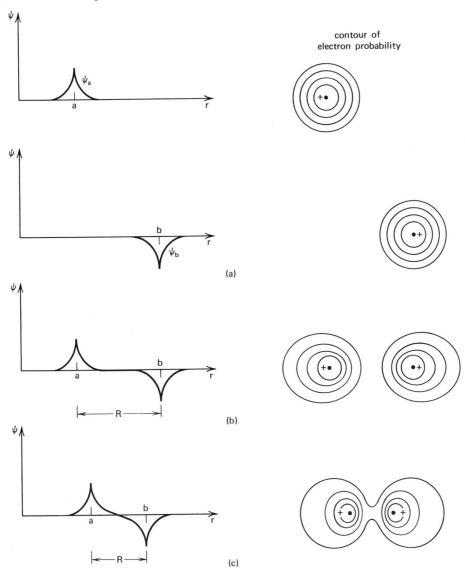

FIGURE 5-4 Electron wave function in the presence of two protons (antisymmetric combination) (*a*) at large separation, (*b*) at intermediate separation, and (*c*) at separation comparable to the wave function spread.

The energy of rotation E_{rot} for a body can be written

$$E_{rot} = \frac{I\,\omega^2}{2}$$

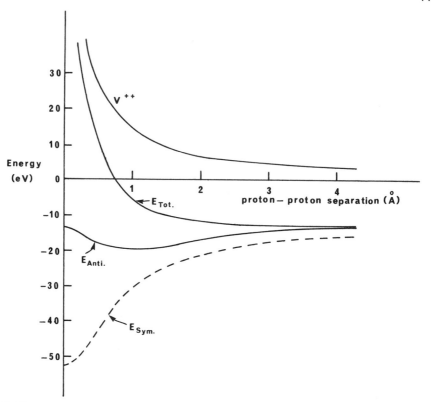

FIGURE 5-5 Electronic, proton repulsion and total energy in H_2^+ as a function of proton-proton separation for antisymmetric electron wave function. The symmetric wave function electronic energy is also shown. Figures 5-2 through 5-5 are adapted from *Concepts of Modern Physics* by A. Beiser, Copyright © 1973, McGraw-Hill, Inc. Used with permission of McGraw-Hill Book Company.

where I is the moment of inertia of the body about the rotational axis, and ω is the angular velocity of the rotating body. To make a "quantum" modification of this energy (impose a quantizing condition) the energy needs only to be written out explicitly in terms of angular momentum, L, which for the rotating body is

$$L = I\omega$$

The energy can be rewritten as

$$E_{rot} = \frac{I}{I} \times \frac{I\,\omega^2}{2} = \frac{(I\omega)^2}{2I} = \frac{L^2}{2I}$$

and the quantum condition becomes (by analogy to the previous treatment)

$$L = \sqrt{\ell(\ell+1)}\hbar$$

where ℓ is any positive non-zero integer and therefore

$$E_{rot} = \frac{\ell(\ell+1)\hbar^2}{2I}$$

where now ℓ represents a *rotational* angular momentum quantum number. What we are really interested in finding out at this point is the relative magnitude of this *rotational*, quantized, molecular energy to the "electronic" energy—the energy stored in the motion of the electrons about their individual nuclei. The rotational energy is of the order of

$$E_{rot} \cong \frac{\hbar^2}{2I}$$

We need to know what the magnitude of I is. For a diatomic molecule, say hydrogen, separated by an atomic spacing we can take I as the moment of inertia of a "dumbbell" consisting of two masses M separated by a distance R;

$$R \cong 1\text{Å} = 10^{-10} \text{ m}$$

$$M = M_H \cong 2 \times 10^{-27} \text{ kg}$$

$$I \cong \tfrac{1}{2} MR^2$$

for rotation about the midpoint of the molecule. Hence

$$I \cong 1 \times 10^{-27} \times (10^{-10})^2 = 1 \times 10^{-47} \text{ kg m}^2$$

so that

$$E_{rot} \cong \frac{(1 \times 10^{-34})^2}{2 \times 10^{-47}} = \frac{10^{-21}}{2} \text{ J} = 3 \times 10^{-3} \text{ eV}$$

The energy of rotation for a molecule is very small compared to the energies involved in electronic motion.

The molecule can also act like a spring and vibrate. Classically, masses connected by a spring oscillate with characteristic frequencies typified by ν_{osc}, where

$$\nu_{osc} = \frac{1}{2\pi} \sqrt{\frac{k}{M}}$$

where M is the mass* of the vibrating system and k is the spring constant. Quantum mechanically this can be treated by including an oscillator potential term of the form

$$V_{osc} \cong \frac{kx^2}{2}$$

* Accurately, it should be the reduced mass.

in the Schrödinger equation where again k is the spring constant and x is the displacement from the equilibrium position. The quantum-mechanical treatment of this problem† leads to the result

$$E_{vib} = (v + \tfrac{1}{2})\, h\nu_{osc}$$

where v is the *vibrational* quantum number and can be any non-negative integer. Once again we can estimate the order of magnitude for this vibrational energy

$$E_{vib} \cong h\nu_{osc} = \hbar \sqrt{\frac{k}{M}}$$

We now need some idea of the magnitude of the spring constant. This can be estimated from the shape of the total energy curve for the hydrogen model by "fitting" it to a harmonic oscillator potential. If the molecular "well" becomes ~ 5 eV deep in $\sim 1\text{Å}$ (the binding energy of molecular hydrogen), we can estimate the value of the "spring constant" as

$$k = \frac{2V}{x^2} \cong \frac{2 \times 5\text{ eV}}{(1\text{Å})^2}$$

$$= \frac{10 \times 1.6 \times 10^{-19}\text{J}}{10^{-20}\text{ m}^2}$$

$$= 160\,\frac{Nt}{m}$$

Estimate of spring constant from TE curve

From the width and depth of well we can estimate k

Assume a simple harmonic oscillator shape for the well ($F = -kx$) so that $PE = \frac{kx^2}{2}$

From the known well shape we estimate $W \cong 5\,eV$ & $x \cong 1\,\text{Å}$

Hence $k = \frac{2W}{x^2} = \frac{10\,eV}{10^{-20}m^2}$

$= \frac{16 \times 10^{-19}\,J}{10^{-20}\,m^2}$

$\to 160\,\frac{Nt}{m}$

and see what that gives for the spacing of adjacent vibrational levels;

$$E_n - E_{n-1} \cong \hbar \sqrt{\frac{160}{10^{-27}}}$$

$$= 10^{-34}\sqrt{1.6 \times 10^{+2}\,10^{+27}}$$

$$= 4 \times 10^{-20}\,J \cong 0.2\text{ eV}$$

† See Further Reading, Norwood, Sec. 8-6.

Of course this is only an order-of-magnitude calculation, but we can quickly see that the energy of vibration is much greater than that of rotation, and less than electronic energies. We therefore expect the total quantum picture of a molecule to be a super-position of electronic states of motion, vibrational states of motion, and rotational states of motion, as indicated in Fig. 5-6. For each electronic quantum state there will be a set of vibrational quantum states, and for each of these vibrational-electronic states will be a set of rotational states.

5.3 RADIATION ABSORPTION The energy level structure picture that we have developed applies basically to isolated molecules only. It is not unreasonable, how-ever, to ask whether there are other circumstances where a similar general picture might apply, although not necessarily in all details. The case for which our picture

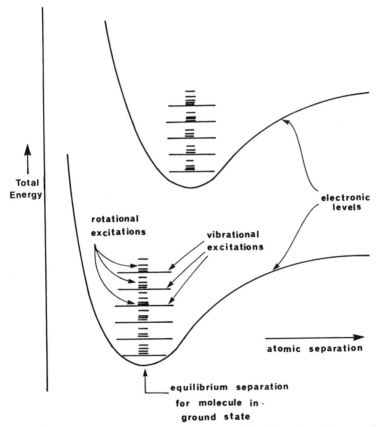

FIGURE 5-6 Total energy of a diatomic molecule as a function of atomic separation, also indicating rotational and vibrational energy levels.

most certainly ought to apply is a molecular gas at not too high a pressure or density so that we are dealing mainly with isolated molecules.

We might also expect similar properties for a *liquid*. Here certainly there are many strong perturbing influences from other atoms on individual atoms which are to form molecules, or multiple molecules. As they are not rigidly held in place and can execute vibration and to a much lesser extent rotational motion, similar types of optical behavior to gases might be expected. A *solid* would almost certainly not be expected to exhibit similar properties in general, since individual atoms are held very strongly in regular arrays due to the influence of large numbers of other atoms. However some similarities do exist, but a different treatment from the above is necessary to understand their causes. We will therefore look at the properties of our model of a molecule and compare them with what happens, mainly in molecular gases.

If we deal with a room temperature gas, we know that the thermal energy of the gas is capable of causing excitations in the molecules, and on the average an excitation of ≈ 0.04 eV might be expected in an individual molecule. Consequently molecules at room temperature are most likely in excited rotational states ($E_{rot} \cong 0.001$ eV) and, remembering the shape of the Maxwell-Boltzmann distribution, quite possibly in a state of vibrational excitation, but almost certainly not in a state of electronic excitation.

What kind of light—electromagnetic radiation—can such an assembly of molecules absorb? We will take as an example H_2. For very low frequencies (heat and infrared), there will most certainly be large numbers of molecules in just the right state to absorb this low-energy radiation and be excited to a higher rotational-vibrational state. Hence the gas ought to appear *opaque* to all radiation up to an energy of (in case of H_2) $\sim$ 4 eV which is the binding energy of the molecule. However, we remember that the electromagnetic radiation carries angular momentum as well as energy, and both must be conserved in a transition. Since the radiation mainly carries one unit of angular momentum, this severely restricts the possible transitions for the higher photon energies. (In fact this requirement restricts the change in vibrational quantum number, v, to one,* so that for photon energies much greater than about 0.1 eV no absorption can occur.) The gas becomes transparent. This continues until the photons have enough energy to excite the molecule into a different electronic state, which will occur at around 10 eV for hydrogen. Then once again by a combination of electronic, rotational, and vibrational excitation the radiation will be absorbed readily; the gas will become opaque to high energy radiation.

The general picture of gas opacity will look something like that shown in Fig. 5-7.

The shape of the opacity vs frequency curve will probably not be smooth, but will depend in detail on the individual energy levels available and temperature and pressure. However the general outline shown should be followed. For *very* low frequen-

* See Further Reading, Beiser, sec. 8-9.

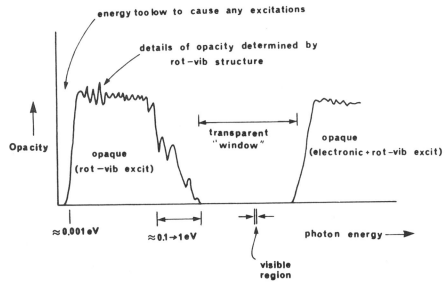

FIGURE 5-7 Expected opacity of diatomic molecular gas as a function of photon energy.

cies, $E_{photon} < E_{rot}$, the radiation should again not be absorbed and the gas will be transparent.

We have used the hydrogen molecule to give us an idea of the basic characteristics of a molecular gas, and the energy scales involved. It is of interest to try to see how these will change if the mass of the material is changed. In general we might expect the rotational and vibrational level spacing to compress since

$$E_{rot} = \frac{l(l+1)\hbar}{M R^2}$$

and the *size* of the molecule remains roughly the same, but the mass increases. Similarly the vibrational energies

$$E_{vib} = (v+1/2) \hbar \sqrt{\frac{k}{M}}$$

are controlled by the mass, but not as strongly. (We expect the spring constant, k to remain *roughly* constant.) Consequently the lower edge of the transparent window should move slowly down in energy. The upper edge of the "window" is determined by the energy for electronic excitation. This is a difficult quantity to calculate in the general case, but we can use the Bohr model to estimate its systematic variation for simple hydrogenlike atoms. The electronic excitation energy, E_{el} should vary roughly as

$$E_{el} \approx -(Z_{eff})^2 \, E_0 \left(\frac{1}{n^2} - \frac{1}{(n+1)^2} \right)$$

where, as discussed before, we need only consider the outermost electron so $Z_{eff} \cong$ 1 and does not change. The value of the major quantum number will generally *increase* with the mass of the atom, as more electrons are added, so that one can expect a general *decrease* (though not in a smooth fashion) in the upper edge of the window. In fact for heavier elements it can be expected in general that the transparent "window" will disappear altogether. For instance the electronic excitation energy for Na would occur at

$$E_{el} \simeq E_0 \left(\frac{1}{3^2} - \frac{1}{4^2} \right) = 0.04 \, E_0 = 0.5 \, eV$$

which is comparable to the expected position for the window lower edge. The window will have been pushed below the visible region (assuming we could produce a gas of molecular sodium!)

5.4 FLUORESCENCE   Let us take a gas of intermediate Z (say nitrogen) so that the electronic excitations are a few eV or less and shine high frequency radiation into the gas so that some of it is absorbed by the molecules as shown in Fig. 5-8. What happens to that absorbed energy? It is later reradiated at random times and in random directions. If one were to look at right angles to the incident light as the gas is being irradiated, the gas itself would be seen to be re-emitting the previously absorbed radiation. This process is called fluorescence, and will have a high frequency spectrum characteristic of the energy level structure of the particular gas that is fluorescing. It is also basically responsible for the blue color of the sky. In this case the fluorescing levels are in the ultraviolet, so that they cannot be seen directly. However because

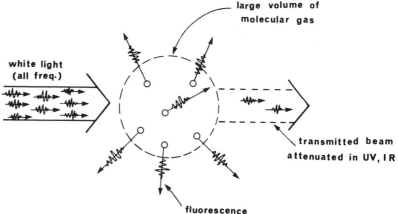

FIGURE 5-8 Fluorescence of molecular gas.

the excited states exist for a finite time before de-exciting, the basic standing wave nature of the state requires that there also be a finite spread to the actual value of the excited state energy of individual molecules (the uncertainty principle). This means that photons with sufficiently similar energy will be absorbed and subsequently emitted even though they do not exactly match the main excited state energy. In this case the cross section for the process is strongly energy dependent, increasing as the fourth power of the photon energy as the main excitation energy is approached. This process has also been described classically and is called Rayleigh scattering, although the "scattering" process is actually an absorption-emission sequence.

5.5 FREQUENCY SHIFTING   In the foregoing we assumed that the frequency of the re-emitted light was essentially the same as that of the incident photon. However the light is not re-emitted instantaneously; typical excited state lifetimes are of the order of 10^{-8} s. During that time it is possible for the excited molecule to collide with another molecule, and share part of its rotational-vibrational energy (Fig. 5-9).

When the molecule finally de-excites to the ground state it will do so by emitting a photon of *lower* frequency than the one that was initially absorbed. This process occurs naturally but it can be enhanced by introducing a small number of molecules that have a lower electronic excitation energy than the main material. The introduced species is called a frequency shifter or a luminescent center. The basic usefulness lies in the fact that these frequency shifters change the frequency of radiation from one that is readily absorbed by the majority of molecules to one for which most are "transparent." This is shown in Fig. 5-10. Although we have talked in terms of molecular gases more practical applications of frequency shifting are found in solids, such as electroluminescent materials and scintillation detectors.

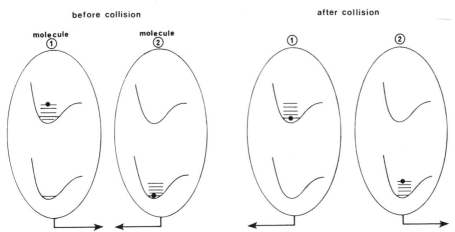

FIGURE 5-9 Collision between molecules, resulting in redistribution of rotational-vibrational excitation energy.

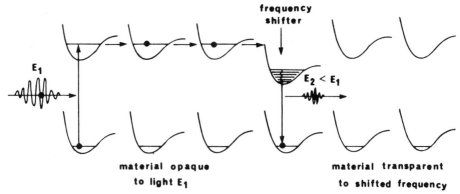

FIGURE 5-10 Occurrence of frequency shifting of absorbed radiation in a molecular gas.

5.6 PHOSPHORS   The length of time that a molecule will remain in an excited electronic state is determined by the detailed quantum mechanical properties of the excited state, the ground state, and the electromagnetic wave associated with the emitted photon. However if the de-excitation transition has to occur by one of the less likely forms of radiation (where the photon has to carry $\ell > 1$, for instance), the time required for this to happen on the average will become much greater than the typical 10^{-8} s for "normal" transitions. The actual value can vary by many orders of magnitude and can become as long as several seconds before de-excitation occurs. Such a situation readily can be imagined for the arrangement of electronic states shown in Fig. 5-11.

where the molecule is excited up to one particular electronic state, then through a radiationless collision changes to an electronic state that is different by more than one

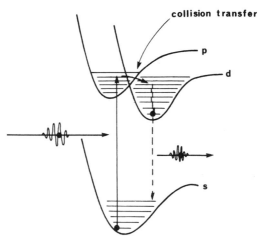

FIGURE 5-11 Phosphorescence in a molecule. In this example the re-emission of a photon is delayed since the molecule has been put into a long lived configuration by means of a collision with another molecule just after absorbing the incident photon.

unit of angular momentum from the ground state, but which can only decay to the ground state. The decay will occur, but after a long time (compared to the 10^{-8} s characteristic lifetime).

5.7 STIMULATED EMISSION So far the picture we have of the interaction of photons and atoms or molecules is that if a photon with an energy E "collides" with an atom or molecule that has an excited state of just the right energy and angular momentum the photon will be absorbed, leaving the atom in the excited state. At some later random time, as discussed before, this system will spontaneously emit a photon of the same energy (we will ignore the possibility of other energy levels in the system).

However this is not all that can happen; there is a second possible process called stimulated emission (Fig. 5-12). If, while the system is in the excited state, an additional photon arrives with exactly the same energy as was absorbed previously, the system will be induced to de-excite, emitting a photon at the time the incident photon

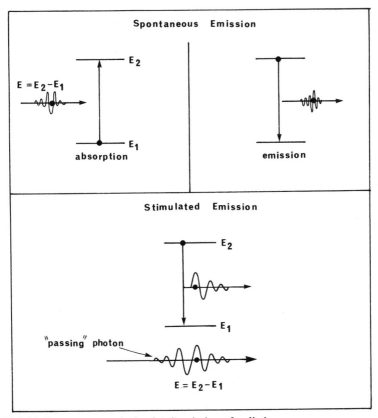

FIGURE 5-12 Spontaneous and stimulated emission of radiation.

arrives. The presence of the incident photon effectively forces the system to de-excite at the precise time of its arrival, rather than at some random time which would otherwise happen.

Such an effect is not entirely unexpected and can be roughly pictured classically by considering the interaction of a traveling transverse electric field and a charged dipole which has a resonant frequency equal to the frequency of the traveling wave.

As illustrated in Fig. 5-13, the oscillating electric field will cause the charge dipole to oscillate at the same frequency and in phase with the electric field. However the oscillating dipole now constitutes an accelerating charge distribution, and will itself generate a traveling transverse electric (electromagnetic) wave which is at the same frequency and has the same phase as the inducing field.

This is in principle what is happening in the case of the atomic stimulated emission, although the analogy should not be carried too far. From this picture it is quite easy to understand some of the basic properties of stimulated emission radiation:

> (a) temporal coherence of the induced and inducing radiation;
> (b) increased probability of stimulated emission with increasing flux of incident photons.

However other aspects can be understood only in terms of the quantum properties of the stimulated emission, such as the fact that it is also spatially coherent (the emitted photon goes in the same direction as the inducing photon).

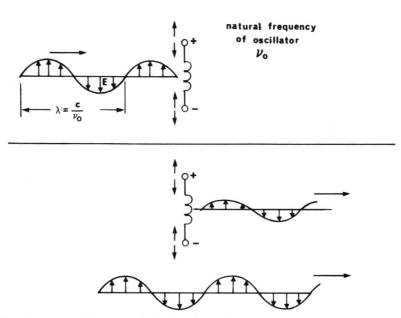

FIGURE 5-13 Classical analog of stimulated emission.

FURTHER READING   A. Beiser, *Concepts of Modern Physics, 2nd ed.*, McGraw-Hill, 1973, Ch. 8.

S. C. Curran, *Luminescence and the Scintillation Counter,* Academic Press, 1953.

E. Fenyves and O. Haman, *The Physical Principles of Nuclear Radiation Measurements,* Academic Press, 1969, sec. 3-4.

H. P. Kallmann and G. M. Spruch, eds., *Luminescence of Organic and Inorganic Materials,* Wiley, 1962.

R. B. Leighton, *Principles of Modern Physics,* McGraw-Hill, 1959, Ch. 9.

J. Norwood, *Twentieth Century Physics,* Prentice Hall, 1976, sec. 8-6.

B. Rossi, *Optics,* Addison-Wesley, 1957, sec. 8-8.

E. Schram, *Organic Scintillation Detectors,* Elsevier 1963, Ch. 2.

J. C. Slater and N. H. Frank, *Electromagnetism,* McGraw-Hill, 1947, sec. 12-6.

P. A. Tipler, *Modern Physics,* Worth Publishers, 1978, Ch. 7.

PROBLEMS   1. Although the molecule He_2 is unstable and does not occur, the molecular ion He_2^+ is stable and has a bond energy about equal to that of H_2^+. Explain this observation.

2. The picture of H_2^+ suggests that the electron spends most of its time between the two protons. How much binding would result in a classical description of this situation (use the observed proton-proton spacing)? What does your result imply?

3. Explain, using the Pauli principle, why the electron spins in a ground-state hydrogen molecule must be antialigned. Why must the spatial wave function be antisymmetric if their spins become aligned?

4. From the treatment of molecular hydrogen, decide what elements should not be able to form molecules. Give the reasons for your answers.

5. Using the Maxwell-Boltzmann distribution, estimate the relative number of hydrogen molecules in the first three vibrational levels at 5000 K. How are rotational levels affected at this temperature?

6. At what temperature would the average kinetic energy of hydrogen molecules be equal to their binding energy?

7. You are to design a high-efficiency insulating window. What kind of gas should you insert between the two glass panes? Why?

8. In what way, if any, do you expect atmospheric pressure to affect the absorption properties of a molecular gas?

9. In an attempt to determine the stability of boron nitride in a reactor environment, you are trying to detect a change in the $^{14}N/^{15}N$ ratio. The possible ways to detect the change appear to be X-ray analysis of the solid, or absorption spectroscopy of the nitrogen when liberated to form N_2 molecules. Use the Bohr model to estimate which of three nitrogen characteristics—electronic, rotational, or vibrational excitations—

will be more sensitive to nuclear mass changes. What is the minimum fractional frequency resolution you will require?

10. Estimate what the upper frequency limit to long-distance ratio frequency transmission would be if the atmosphere was mostly Cl_2.

11. Calculate what the frequency of radiation should be when a molecule of HI (hydrogen iodide) de-excites from the $\ell = 5$ to $\ell = 4$ state (ℓ is the rotational quantum number). Due to the high mass of the I the moment of inertia can be taken to be $I = MR^2$. Assume $R = 1\text{Å}$. Would you expect this to be a sensitive way to test for, say, HBr in the presence of HI?

12. Calculate the correct moment of inertia of molecular hydrogen in its ground state.

13. Estimate what the probability is for a molecule to undergo a collision while in an electronic excited state in a gas at NTP. Use a characteristic electronic de-excitation time of 10^{-8} s, and the concept of mean free path (see question 2-10).

14. Reduced power output from a heavy water production facility suggest that there may be significant losses of heavy water through leakage to nearby light water cooling lines. It has been decided to test for the presence of deuterium by electrolyzing the cooling water and monitoring the absorption spectrum associated with either rotational or vibrational excitations of the evolved gas. Which of the two types of excitation would provide the greater fractional frequency shift for detecting deuterium? Calculate (to $\sim 10\%$) the lowest energy transition which could be used to indicate the presence of deuterium.

15. It has been observed that the rotational energy level spacing of diatomic molecules begins to deviate from the expected $\ell(\ell+1)$ behavior when high angular momentum states are reached, and the effect has been ascribed to stretching of the molecule. Using classical arguments estimate how much deviation should occur due to this effect as a function of ℓ.

16. The spring constant for a single diatomic molecule was estimated in the text. Assuming all atom-atom bonds are approximately the same, how would you expect the molecular spring constant to be related to spring constants for solids? What approximate scale factor would you expect between the two?

17. Determine what elements should have low-lying states with a large difference in ℓ, and therefore might exhibit phosphorescence, by using the shell filling order indicated by Table 4-1.

18. Significant clustering of molecules is found for clusters up to a size scale of 100 Å. How would you expect the relative probability of light scattering from a cluster to depend on N, the number of molecules in the cluster? Should your answer depend on the wavelength of the light? Why?

SIX

APPLIED ATOMIC AND MOLECULAR PHYSICS: LASERS

6.1 LIGHT AMPLIFICATION BY STIMULATED EMISSION   The fact that atomic or molecular systems can be made to de-excite by stimulated emission means that it should be possible to amplify a light signal under the appropriate conditions. These appropriate conditions can be recognized by considering what will happen to photons passing through a collection of atoms or molecules which can be characterized for simplicity by a two-level structure where the energy difference is exactly the same as the incident photon energy. When the photons and atoms or molecules interact, one of two processes* can occur: if the quantum system is in the lower state, photon absorption will take place; if the system is in the upper state, then stimulated emission will take place, and the number of photons at that frequency will be increased—an amplification process occurs. Quite clearly the necessary condition for light amplification to take place is that the probability of stimulated emission exceed the probability of absorption. This can occur only if there is a *greater number* of atoms or molecules in the upper state than in the lower state. Such a condition is called a *population inversion* or *negative temperature*. When this condition is attained the arrival of a single photon is capable of triggering an avalanche of coherent (in-phase) photons, producing radiation many orders of magnitude more intense than the initial radiation.

The coherence of the emitted photons is very important to the attainment of a high intensity light output (a high system gain). This can be seen readily by comparing the light intensity at some point from a coherent and an incoherent light source as shown in Fig. 6-1. For a *coherent* source (made of many radiating atoms) all these radiating atoms are in phase.

This means that at any particular time the amplitude from the individual atoms are in phase on the screen and we obtain for the total amplitude of the light and therefore the total intensity, I_{coh};

* We are ignoring spontaneous emission.

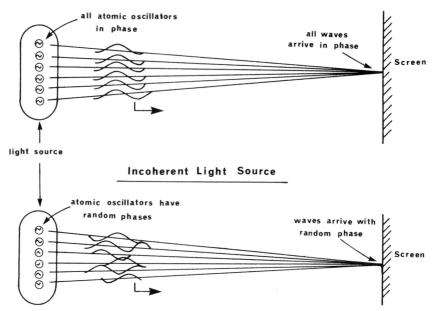

FIGURE 6-1 Light output from a coherent and an incoherent source.

$$I_{coh} = A_{tot}^2 = (a_1 + a_2 + a_3 + \ldots\ldots a_y)^2$$

Since we assume that the individual radiators are all the same type of atom, the amplitudes are all characterized by $a_i = a_0 \cos \phi$, where ϕ is the phase of the wave from an individual atom at the screen. Thus

$$I_{coh} = (Na_0 \cos\phi)^2 = N^2 a_0^2 \cos^2\phi$$

that is, the total intensity increases as the *square* of N, the number of radiators.

For an *incoherent* source this is not the case. To illustrate this take two atomic radiators with some arbitrary relative phase; the amplitudes add according to

$$A = a_0 \cos \phi_1 + a_0 \cos \phi_2$$

For convenience we will measure all phases relative to ϕ_1. This gives us

$$I_{incoh} = A^2 = (a_0 + a_0 \cos \phi_2)^2$$
$$= a_0^2 + a_0^2 \cos^2 \phi_2 + a_0^2 \cos\phi_2$$

For a very large number of incoherent radiators this gives

$$I_{incoh} = \Sigma_i \, a_0^2 \cos^2 \phi_i + \Sigma_{ij} \, a_0^2 \cos \phi_i \cos \phi_j$$

If there is indeed a very large number of radiators and if the times that they start to emit their radiation are entirely uncorrelated, then all values of ϕ_i and ϕ_j will occur with equal probability in the above summation. In that case, the summations are equivalent to an average over all possible angles. Since $\int_0^\pi \cos^2 \phi \, d\phi = \pi/2$ and $\int_0^\pi \cos\phi d\phi = 0$ this gives for the incoherent intensity

$$I_{incoh} = Na_0^2 + 0$$

$$= Na_0^2$$

The intensity for an *incoherent* source increases only linearly with the number of radiators. Basically, part of the possible light intensity is lost due to interference.

In practice laser amplification normally is obtained by exciting atoms or molecules into long-lived levels to ensure a population inversion and hence stimulated emission. That populating a long-lived state is necessary, or at least highly desirable, can be recognized quickly by estimating the power requirements necessary to achieve a population inversion for a "normal" lifetime system. Roughly, the requirement should be that there is enough energy injected into the system to put at least half of the atoms or molecules in an excited state on a time scale no greater than the characteristic lifetime. This gives the following estimate

$$\text{Minimum power} = \frac{E \text{ per excitation} \times \text{# of atoms present} \times \text{fraction excited}}{\text{average decay time}}$$

$$E \text{ per excitation} \sim 2 \, eV$$
$$\text{number of atoms} \sim 6 \times 10^{23}$$
$$\text{fraction excited} \qquad 50\%$$
$$\text{decay time} \sim 10^{-8} \, sec$$

$$\text{Power} = \frac{2 \, eV \times 1.6 \times 10^{-19} \frac{J}{eV} \times 6 \times 10^{23} \times 1/2}{10^{-8}}$$

$$= 10^{13} \, watts \; !!$$

Clearly, if a population inversion is to be obtained simply by "pumping" atoms or molecules out of the ground state, large amounts of power will be required, or a very long-lived excitation compared to the typical value of 10^{-8} s we used in our calculation must be available, or both. In fact population inversions are achieved in a number of ways and we will look at several in the next section.

The basic characteristics of any laser system, outlined in Fig. 6-2, will consist of some form of lasing material into which energy is pumped to provide a population inversion so that stimulated emission will occur. The efficiency of amplification is

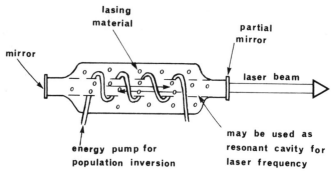

FIGURE 6-2 Basic laser system.

normally enhanced by using a pair of very accurately aligned mirrors so that the light beam passes many times through the lasing material. The laser beam is extracted either by using a partly transmitting mirror at one end (in order to obtain a continuous beam) or by using a mirror whose reflectivity can be "shut off" rapidly once the laser beam has reached the desired intensity in order to let the beam escape from the laser cavity.

The resulting light beam is highly monochromatic, since only the right energy photons will be amplified, highly collimated, due both to the many passes through the mirror system and to the inherent spatial coherence; and extremely intense due to the temporal coherence of the amplification process.

6.2 TYPES OF LASER OPERATION   There are many types of lasers, and it is worthwhile to look at the operation and characteristics of some of them. Perhaps the most common is the He-Ne gas laser. In this device helium atoms are excited to long-lived states by collisions with energetic electrons. These electrons are produced by ionization of neutral atoms caused by applying either high DC or RF electric fields and then accelerating the electrons to high velocites. The He excitation energies match very closely the excitations in Ne, shown in Fig. 6-3 labeled s, and the excitation energy of the atoms is transferred to the Ne atoms by atomic collisions. When the excited Ne levels are produced in this way, an automatic population inversion is created, since the lower energy p states are not occupied at all, and hence stimulated emission can take place for the three sets of $s \rightarrow p$ transitions shown in Fig. 6-3. The subsequent $p \rightarrow s$ transition will not lase, since the lower s states are long-lived and very quickly become more populated than the feeding levels. (The spontaneous decay lifetime for these lower states is so long that the normal mode of de-excitation is by collisions with the walls of the laser.) This type of laser is relatively inefficient (typically about 50 mW out for a 50 W input) due at least in part to the fact that only a rather small fraction of the de-excitation energy is involved in the laser transitions (i.e., $\sim 1 \rightarrow 2$ eV out of 20). The main role of He in this laser is to increase the pumping efficiency, since it has no low-lying levels, unlike Ne.

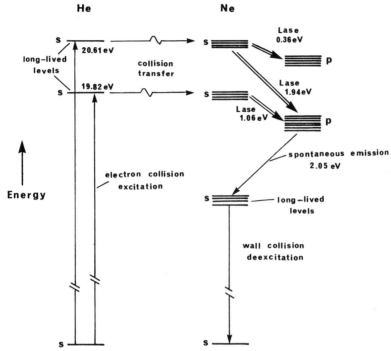

FIGURE 6-3 Energy levels for He-Ne laser.

A second type of lasing system, characterized by the ruby laser, operates by optical excitation of "impurity" ions (for ruby, Cr^{3+}) embedded in a transparent crystal or glass (for ruby, Al_2O_3). The relevant energy levels are shown in Fig. 6-4.

The Cr^{3+} ions are excited into the levels labeled F_1 and F_2 by absorbing photons produced by an intense light source. (A frequently used light source is a Xenon lamp. Xenon atoms are excited by electron impact as in the He-Ne laser.) The excited Cr^{3+} ions very rapidly de-excite via interactions with the Al_2O_3 lattice to the state labeled E, which is long-lived ($\tau \cong 4$ ms). If the pumping light source is intense enough, more than 50 percent of the Cr^{3+} ions can be pumped into the long-lived state, and lasing will occur. Considerable power is required, however; as much as a megawatt power input to the Xenon lamp for a millisecond or so can be used, and a laser output power of the order of 20 kilowatts can then be obtained. This laser is much more efficient than the He-Ne laser, since a large fraction of the excitation energy is converted to laser light, but it still represents an overall efficiency of at best a few percent. Because of the very high input power requirements, plus the high heat inputs associated with this power level, this type of laser is only practicable as a pulsed source when run at high power levels.

A third type of operation is found in the CO_2 laser, which utilizes collision pumping to populate vibrational states. The CO_2 molecule is capable of three different modes

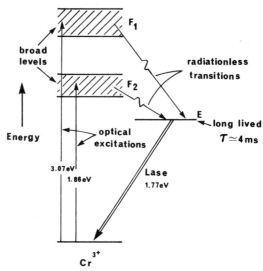

FIGURE 6-4 Energy levels for Cr^{2+} ions in ruby laser. The levels marked F_1 and F_2 are broad due to the influence of the Al_2O_3 lattice.

of vibrations. These set up the somewhat complicated spectrum shown in Fig. 6-5; the individual levels are labeled by the number of vibrational quanta of each mode. The CO_2 is excited to a long-lived vibrational state ($\tau \cong 2$ ms) by electron impact; a small amount of N_2 is often added to the CO_2 gas to improve the pumping efficiency, as in the He-Ne laser. Although the state is long lived, a population inversion with

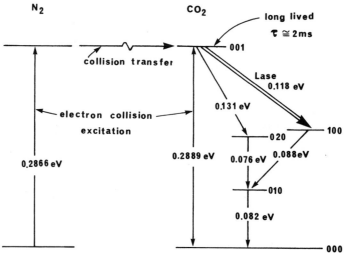

FIGURE 6-5 Energy levels for a CO_2 laser. Pumping may be aided by the presence of N_2. CO_2 levels are labelled by the three separate vibration quantum numbers.

respect to the ground state does not need to be attained for this device to lase, instead the laser cavity is tuned to resonate at the frequency that corresponds to the radiation from the 0.118 eV transition to the state labeled 100, so that this becomes the laser transition and a population inversion *relative to the 100 state* can be achieved at much lower power levels than would otherwise be necessary. Because of the higher efficiency of transferring energy via electron collisions at sub-electron volt energies much greater pumping efficiencies can be achieved. Also, the CO_2 level scheme is such that approximately half of the excitation energy can be converted into laser energy. In fact, together with the very high pumping efficiency, a tremendously high overall efficiency of approximately 30 percent has been obtained with this type of laser. It is also capable of delivering up to about 100 W of continuous power and on the order or 3×10^{10} W in very short ($\sim 10^{-9}$ s) bursts! The high power capability of this laser is being developed actively.

A more recent type of laser makes use of a different molecular property to produce a population inversion. It is called an *excimer* laser (Fig. 6-6) and uses atoms that normally do not form stable molecules, such as the rare gases. Because of their closed-shell electronic structure, the only interaction such atoms have with each other when they are in their ground states is a repulsive one. However, when one of them is in an excited state it is possible for two atoms to bind together to form an unstable molecule, since now both attractive and repulsive interactions between the atoms are

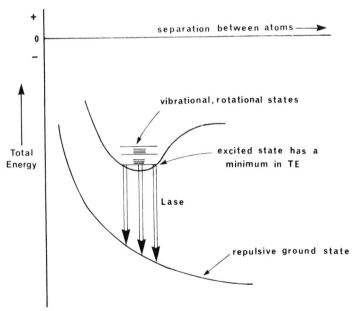

FIGURE 6-6 Energy levels for an excimer laser, showing repulsive ground state, minimum in excited state total energy which allows the formation of an excimer, and the variation de-excitation photon energy.

possible. This type of unstable, excited state molecule is called an *excimer* and is denoted by placing an asterisk after the chemical element symbol. Once such a compound is formed, there is automatically a population inversion since any particular pair of ground-state atoms are extremely unlikely to be at the right separation to absorb the excimer state deexcitation radiation.

An example of the excimer laser is the Xe_2^* laser. Excited Xe states are produced by electron impact excitation, using a pulsed electron beam. By maintaining a high pressure in the Xe gas ($\gtrsim 10$ atm) there is a high enough chance of Xe-Xe collisions while one of the pair is in an excited state to form a significant number of excimers. This results in a population inversion, and when the gas is placed in an optical cavity of the appropriate dimension a usable laser beam can be produced. The de-excitation photon energy is approximately 7.1 eV, but due to an unusual combination of level properties for this transition (Fig. 6-6) the energy is not fixed uniquely. Because of the vibrational property of the excimer state, there is a range of atom-atom separations that can exist at any time, and in particular at the moment when the excimer state de-excites. However, the energy of the dissociation ground state varies (by approximately 2%) over this same range of separations, so that the de-excitation photon energy can vary by the same amount. This fact can be utilized to produce a continuously tunable laser by adjusting the laser cavity resonant frequency to any value within the range of emitted photon frequencies. The laser transition energy is ~ 20 percent of the total excitation energy so that, together with the relatively high electron beam pumping efficiency ($\gtrsim 80\%$), overall efficiences on the order of 15 percent can be obtained.

6.3 LASER APPLICATIONS; HOLOGRAPHY One development that has been made practicable with the advent of the laser (an intense, coherent light source) is holography. This is a process where an interference pattern between the waves of a reference beam and coherently reflected light from an object is "photographed" so that a full, three-dimensional image of the object can be reconstructed. The basic properties of a holograph can be understood by looking at the pattern of waves from two coherent point sources. Because the two sources are coherent they will produce a standing wave pattern in the region between them. For simplicity let us consider only the points along the crests of constructive interference. These will occur along curves where the phase difference (i.e., difference of pathlength from the two source positions) between the two waves is constant and equal to an integer multiple of 2π. These curves are hyperbolae with the source positions as foci (Fig. 6-7).

If a photographic plate is exposed in the region of this interference pattern, the parts of the film along the surfaces of constructive interference can be developed, but not elsewhere. If, after processing the photographic plate, a point light source is placed at A, part of the light will be intercepted by the developed surfaces, and (because of the properties of hyperbolic reflecting surfaces) will produce a virtual image of the point at B as shown in Fig. 6-8, when the plate is viewed. Since this is equally true

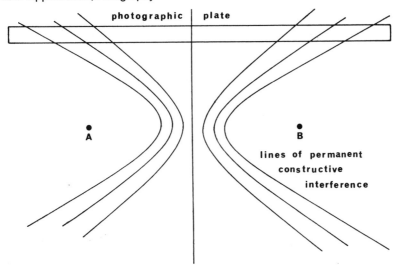

FIGURE 6-7 Lines of constructive interference from two coherent light sources placed at points *A* and *B*. The photographic plate will be developed along the dark lines.

for any choice for point B and for any number of points B, it is possible to record on a photographic plate the interference pattern from a single coherent source at A and the coherently scattered beam from an *extended object* B.

This can be achieved in practice by splitting a laser beam to provide both the reference beam and the coherently scattered light from the extended object to produce the interference pattern which is photographically recorded as indicated in Fig. 6-9.

When the plate is developed, it can be used as before to produce a three-dimensional image of the source, which can be viewed from a range of angles to give different perspective of the original object (Fig. 6-10).

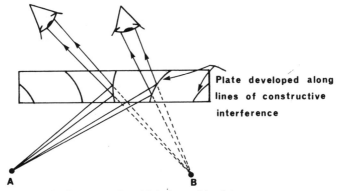

FIGURE 6-8 Reconstruction of point source *B*, with holographic plate.

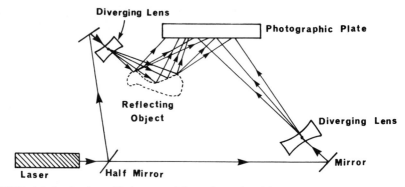

FIGURE 6-9 Production of hologram of three-dimensional image.

HOLOGRAPHIC IMAGE RECONSTRUCTION

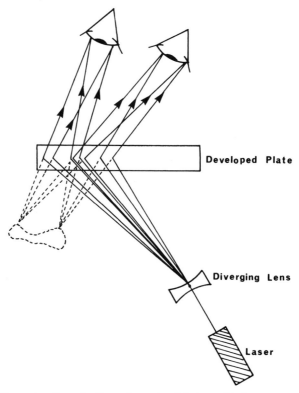

FIGURE 6-10 Reconstruction of a three-dimensional holographic image.

Although the properties of holographic images have produced much interest in themselves, they have valuable applications for observing and measuring very small displacements in objects, such as microscopic strains and distortions. To do this one takes a holographic double exposure of an object. Because each exposure records an interference pattern *for each point* of the object, *both* exposures will produce phase-coherent images, and these will destructively interfere with one another wherever the scatterer position has changed by an integral number of half wavelengths between exposures. Consequently it is possible to observe position changes or distortions of objects on a size scale down to fractions of microns using such a technique. An example of this technique for detecting small distortions is shown in Fig. 6-11.

6.4 ACOUSTIC HOLOGRAPHY   The principle of holography is not limited to coherent light sources, but can be applied using any coherent source of waves whose

FIGURE 6-11 Reconstruction from an interferometric hologram of a dynamically excited turbine blade. From Interferometric Holography by J. P. Waters, in *Holographic Nondestructive Testing,* edited by R. K. Erf. Copyright © 1974 by Academic Press, Inc. Photograph courtesy of the author.

presence can be recorded in some way. A promising technique is acoustic holography. As in optical holography, an acoustic holograph is produced by the interference of a reference beam (Fig. 6-12) with waves from a second, coherent source that is either scattered from or transmitted through the object being examined. The acoustic waves are normally transmitted through a liquid, and their standing wave pattern on the surface of the liquid serves the same purpose as the photographic plate for the optical laser. Because the surface interference pattern cannot be recorded permanently, it must be "viewed" as the pattern is produced. One way that this can be done is by shining a coherent light source (from an optical laser) on the surface. The crests of the liquid standing wave pattern will act in the same manner as the developed parts of an optical hologram, selectively reflecting light when viewed from a non-normal angle. In order to obtain adequate spatial resolution to be sensitive to relatively small details of the object being studied it is necessary to use very high frequency sound waves (in the MHz range). This produces a complication in the optical viewing system because the difference in wavelength used to produce and view the holograph produces a demagnification of the picture, roughly by the ratio of the two wavelengths, so that an optical magnifying system is required.

The absorption and reflection properties of many materials (notably plastics and organic tissues) are very different for ultrahigh frequency sound and light, so that different types of information can be obtained using the two techniques. This is illustrated in Fig. 6-13, where interior recesses of an optically opaque object are sensed by an ultrasonic beam.

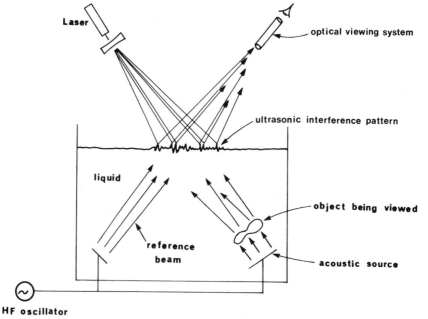

FIGURE 6-12 System for producing and viewing acoustic holographs.

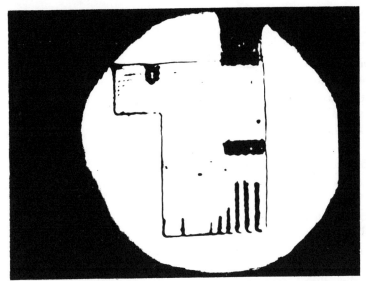

FIGURE 6-13 Acoustical image of an optically opaque plastic block showing the presence of holes drilled in from the bottom and side of the block. The acoustical frequency is 5 MHz. Picture courtesy of Holosonics, Inc., Richland, Washington, USA.

Both lasers and holography are relatively recent technical developments, and are presently undergoing a rapid expansion in both their technical devlopment and possible practical applications. The coherence of the laser lends itself to interferometry and long distance measurement. The high power densities possible has led to precision welding, cutting, and drilling applications as well as small area annealing of metals. Presently holography has been applied to stress and vibration analysis, information storage, image deblurring, and medical diagnostics.

FURTHER READING   M. J. Beesley, *Lasers and their Applications,* Taylor and Francis, 1972.

E. Camatini, ed., *Optical and Acoustical Holography,* Plenum, 1972.

S. S. Charschan, *Lasers in Industry,* Van Nostrand-Reinhold, 1972.

H. J. Caulfield and S. Lu, *The Applications of Holography,* Wiley, 1970.

R. K. Erf, ed., *Holographic Nondestructive Testing,* Academic Press, 1974.

W. E. Kock, *Engineering Applications of Lasers and Holography,* Plenum, 1975.

M. Ross, ed., *Laser Applications, Vols. 1–3,* Academic Press, 1977.

A. E. Siegman, *An Introduction to Lasers and Masers,* McGraw-Hill, 1971.

O. Svelto, *Principles of Lasers,* Plenum, 1976.

PROBLEMS   1. Why should the addition of He in the He-Ne laser and N_2 in the CO_2 laser increase the pumping efficiency? What are the necessary characteristics for such materials used to improve pumping efficiency?

2. Why should the two excited states of He used to pump the He-Ne laser have a longer-than-normal lifetime?

3. Estimate the maximum efficiency theoretically possible for the He-Ne and CO_2 lasers. Give reasons why these efficiencies are unlikely to be achieved in practice.

4. Do you expect the laser temperature to affect the lasing efficiency of a He-Ne laser? a Ruby laser? a CO_2 laser? Estimate the temperature sensitivity of the one that you expect to be most affected by temperature.

5. What effect does the choice of gas pressure have in the operation of an excimer laser?

6. Discuss the expected comparative efficiency of electron impact pumping in He-Ne, CO_2, and Xe_2^* lasers.

7. A laser manufacturing firm is attempting to increase the efficiency of their He-Ne lasers. They are considering the possibility of pumping up the Ne in isolation by electron impact, then mixing it with the Ne to generate a population inversion. Make a rough estimate of what the pumping efficiency could be by comparing the probability of spontaneous de-excitation by the He to the probability of collision with Ne gas. Assume that the time required to mix the two gases is negligible, that any He collision that occurs is with a Ne atom and causes an excitation to the lasing level. Assume the gas is at NTP and that the probability of collision is given by the physical atomic cross section.

8. Estimate what the time history of the upper level of a two level laser system is for a constant pumping rate p_0 to the upper state, which has a mean life τ. Assume that all de-excitation radiation is lost from the system. What other factors would have to be included if this assumption were not made?

9. Estimate what minimum optical pumping power would be required to maintain the Cr^{3+} ions in a 100 g ruby rod *just under* a population inversion (assume all fluorescence radiation is lost from the crystal without being reabsorbed). To how much heating power to the crystal does this correspond?

10. Estimate the improvement in laser pumping efficiency possible using the technique discussed in problem 7.

11. A certain type of continuous output laser is being considered for use in long-range interference measurements. It is found that the light output from this particular type of laser consists of a continuous series of short bursts of light, each burst lasting approximately $0.1 \ \mu s$, despite the fact that the input power is constant. What is likely to be the cause of this unexpected behavior? What is the maximum distance over which you would expect to obtain interference effects (the coherence length) with this laser? What could you do to increase this length?

12. The availability of very high power lasers has led people to consider isotopic separation by means of selectively exciting different isotopes of the same element.

Use the Bohr model to estimate the largest anticipated fractional change in excitation energies for heavy water, zirconium, and uranium. Are these values presently achievable with lasers? What order of magnitude laser power would be required to prepare 1 kg of uranium for separation per hour?

13. Is it possible for two bosons (photons) to exist in exactly the same quantum state? Use the wave function obtained in sec. 4-6 to decide.

14. Would stimulated emission be possible if photons were fermions? Why?

15. Prove that the loci of constant phase between two coherent point light sources produce hyperbolae.

16. What frequency ultrasound must be used if a spatial resolution of 0.1 mm is to be achieved? Assume the ultrasound conducting medium has the same characteristics as water.

SEVEN

THE SOLID STATE
OF MATTER

The quantum properties of atoms and molecules provide the basic information necessary to build up a picture of the properties of solids. There are many interesting, and important, properties of solids that might be considered, but we will restrict ourselves to two, electrical and optical, with the main emphasis on electrical properties.

7.1 BAND THEORY   An atom in a solid has *many* neighbors that will influence its properties, rather than just the one that it has in a diatomic molecule. As we saw (Figs. 5-3, 5-5) interaction between atoms resulted in a modification of the energy levels that "looked" like either an attractive or a repulsive interaction, depending on the detailed nature of the electron wave functions. In determining how an energy level will be changed by the presence of many atoms we will build on the insight gained in studying the H_2^+ molecule. This time we will consider the energy of an electron and a row of four protons. Since the four protons would be hydrogen atoms in reality, we will assume that the proton-proton spacing is approximately the same as we found in the molecular case. As can be seen in Fig. 7-1, there are a number of possible energy levels that the electron might be in, depending on the particular combination of interactions that happen to occur. We can work out the *possible* values of an energy level by listing all the changes in energy its neighbors might induce.

The electron can exist in any one of the levels we have calculated, whose value depends in detail on the interaction with the various neighbors. It is most strongly influenced by the nearest neighbors. As more neighbors are added they are farther away, and have a correspondingly smaller influence. In general as the number of neighbors increases indefinitely an individual electron can have an energy that is limited only to within a range whose envelope is similar in shape to the curves in Figs. 5-3 and 5-5, but somewhat greater in magnitude.

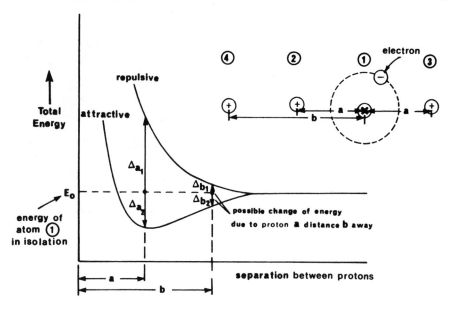

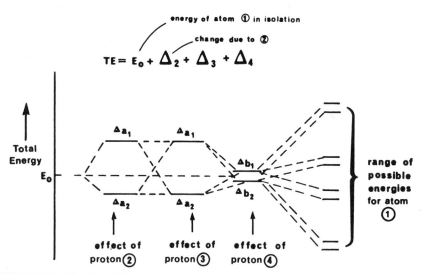

FIGURE 7-1 Modification of hydrogen electron energy due to presence of nearby protons.

It might appear from Fig. 7-1 that an electron in a hydrogen atom can occupy a greater *number* of levels because other atoms have been brought near to it. This is not the case. An electron of hydrogen in its ground state can occupy one of two states; spin up or spin down. An electron in a group of four protons can occupy any of eight states. However, when three more electrons are added to make four hydrogen atoms, four of the states will be occupied, so that once again there are two states available per electron. The effect of bringing atoms together is to *modify the energy* of individual levels, but not increase the total number of levels.

If we make a "solid" from four hydrogen atoms, there will be (two times) four levels spread over some energy range. In a real solid there are $\sim 10^{23}$ atoms so there will be $\sim 10^{23}$ individual levels spread throughout the range, forming a *band* of allowed energies (Fig. 7-2). It is not possible to predict precisely what energy level an individual electron will have—at least to within the band width. The same basic situation applies to the valence electrons of any atom.

7.2 SPATIAL EXTENT OF ELECTRON WAVE FUNCTIONS   From the fact that electron wave functions extend farther away from the nucleus for higher quantum-number energy levels, we would expect the mutual influence between atoms to occur at greater separations for the higher quantum numbers. This is actually what happens, as can be seen for the case of sodium quantum levels shown in Fig. 7-3. At the interatomic separation of solid sodium, the $1s$, $2s$, and $2p$ levels are not affected significantly by the presence of other sodium atoms, but the $3s$, $3p$, and higher levels are strongly influenced and have become bands. Also, there is an energy minimum

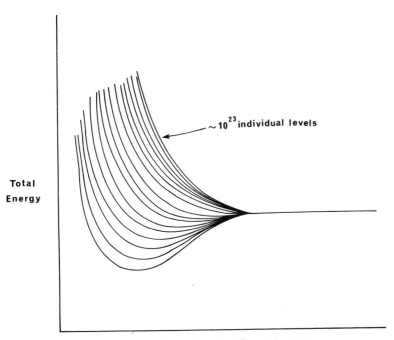

FIGURE 7-2 Formation of energy bands in solids.

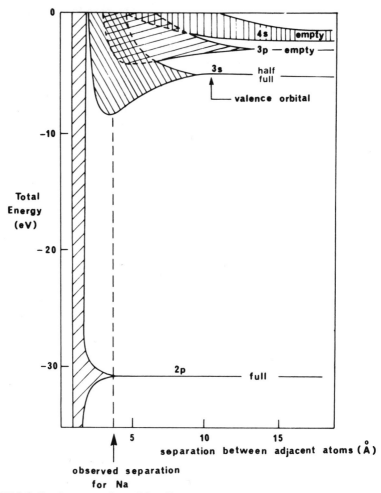

FIGURE 7-3 Band structure for solid sodium.

for the $3s$ electrons that enables a stable solid configuration to exist with a mean nearest-neighbor spacing of $\sim3\text{Å}$.

The fact that the energy of the $3s$ electron in sodium atoms is influenced by neighbors also means that their wave functions overlap appreciably, and it is no longer possible to say that any particular $3s$ electron is uniquely associated with any particular atom. This perhaps can be seen most readily by going back to the Bohr picture of atoms and energy levels; this picture describes the extent of the electron's wave function well enough for our purpose. When the potential well is modified to include the presence of other atoms (Fig. 7-4) we find that the Coulomb potential no longer confines the $3s$ electrons to any particular atom, so they can be anywhere within the solid. This is equivalent to the previous quantum mechanical statement about the overlap of the

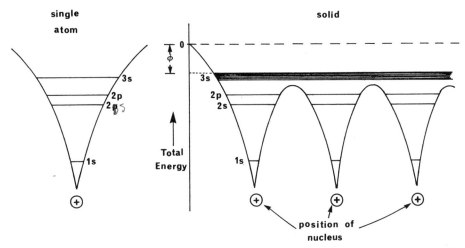

FIGURE 7-4 Bohr atom picture for a solid showing extended range of least-bound electron's orbit.

3s electron wave function. Notice, however, that it is not possible for the electrons to escape freely from the material, since their wave functions do not extend beyond the edge of the material; the electrons are bound in the material by an amount equal to the work function ϕ.

7.3 CONDUCTION PROPERTIES   From the fact that the least-bound electrons are not tied to any particular atom, it should be possible to generate a net flow of electrons through the material. However deciding whether or not this is true requires a more detailed examination.

What is the "ground state" configuration of a solid, that is, its state of *lowest possible total energy*? If we were to very gently put together a collection of atoms in their ground state, removing all excess energy as they are brought together, then every electron would be in the lowest state of total energy allowed by the Pauli principle (Fig. 7-5). In this case there would be *no* kinetic energy in the form of net linear motion.

If there is to be any net linear motion of the electrons it must be due to a total energy in excess of the minimum possible value; it can *only* be achieved by exciting electrons from their lowest state into a higher one. This *excess* energy can then appear as a net linear motion.

Under what circumstances will it be possible to have a net motion of electrons in solids? That is, when can electrical conduction (or a closely related property, thermal conduction*) occur? The answer depends on the occupation of the energy band of the

* Thermal conduction is not exactly the same as electrical conduction since lattic motion (vibrations, etc.) also contributes to the thermal conduction.

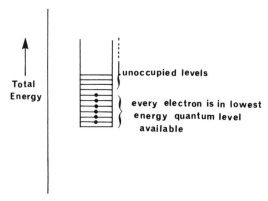

FIGURE 7-5 Schematic diagram of material in minimum total energy state.

electrons. If there are more energy levels in the band than there are electrons to fill them then excitation, and consequently conduction, can occur readily; if, there are as many electrons as levels in a band, conduction will be impossible or at least severely hindered. For example if we consider sodium, then the outermost band is a 3s one; it is possible to put two electrons into any 3s quantum level, and each sodium atom has only one electron in that level. The top of the filled levels in a minimum energy arrangement (''ground state'') only comes to the middle of the band, so it is quite possible to excite an electron to a higher level in the band either thermally or electrically. Consequently sodium is a good electrical and thermal conductor. However one can not always rely on single-atom considerations to predict correctly conduction properties of solids. For instance the above argument would lead to the conclusion that magnesium, which has two electrons in the 3s level (so it is fully occupied) should not be a conductor since the 3s band should be completely full. In fact magnesium *is* a conductor since the energy bands (which will be virtually the same as for sodium) are such that at the observed atomic spacing for magnesium the 3s band overlaps the 3p, 4s, and higher bands, so that unoccupied levels are still available to the least-bound electrons and thermal excitation or excitation by external electric fields can occur readily. Notice that this would *not* be so if, for instance, the interatomic spacing were ~ 8 Å where the 3s and 3p bands become well separated. Electrons could not be excited readily to unoccupied levels and conduction could not occur; magnesium would be an insulator at such an atomic separation.

There are other complicating factors. One would also expect that any material with only a partially filled p shell would always be a conductor. This is not necessarily so. For the case of two electrons in the p shell, such as in silicon, germanium, and carbon in the form of diamond, there is a mixed configuration of the s and p orbitals which produces an especially favored configuration of the outer four electrons, shown in Fig. 7-6 (e.g., the electron-electron Coulomb repulsion is minimized). The s and p wave functions for the electron form what are called hybrid orbitals and result in a restructuring of the bands so that, rather than having a completely filled s band and

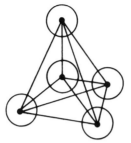

FIGURE 7-6 Tetragonal (diamond) structure. The four outer atoms are at their maximum possible mutual separation in this configuration.

partly filled p band, these materials have one completely filled and one completely empty hybrid s-p band. As can be seen in Fig. 7-7, there is an appreciable energy gap between the filled band and the empty band.

From such considerations materials can be divided into two distinct classes and one that is in between the two as shown in Fig. 7-8. These are conductors (where at least

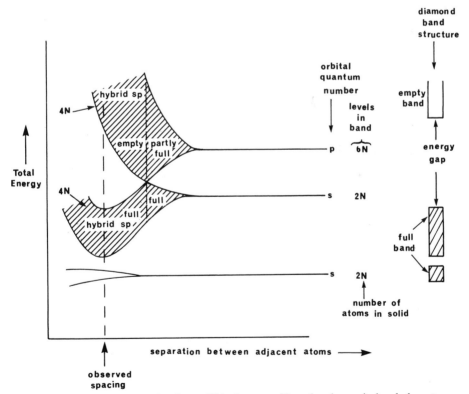

FIGURE 7-7 Band structure for Group IVA elements. Note the change in band character as the p and s bands cross. The observed atomic spacing for diamond is indicated.

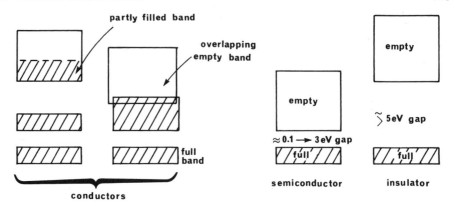

FIGURE 7-8 Band properties of conductors, insulators, and semiconductors.

one electron per atom is freely available for thermal excitation or electric field acceleration), insulators, (where there is a completely filled band of valence electrons and a large energy gap between it and the first available unoccupied state so that virtually no electron excitation can occur), and semiconductors (which also have a filled band, but the gap is sufficiently small that there is a chance that a few electrons can be excited to unoccupied states).

The difference between semiconductors and insulators is only a matter of the magnitude of the energy gap between the filled and empty bands. An insulator is typified by a gap of ~ 5 eV or greater; semiconductors have a gap somewhere on the order of $3 \rightarrow 0.1$ eV. If the gap is much greater than this the probability of excitation to the empty band becomes negligible (as we shall see later); if the gap is much smaller than 0.1 eV the probability of thermal excitation into the empty band becomes very high, and the material acts very much like a conductor. The various valence-4 materials completely span this region. Carbon (diamond) has a gap of 5.47 eV and is an insulator; Si, Ge, and Sn have gaps of 1.12, 0.67, and 0.08 eV respectively and are semiconductors; Pb has virtually no gap and is a conductor (although a relatively poor one).

The highest energy (least bound) band that contains electrons is called the *valence band*. The first band where there are unoccupied levels is called the *conduction band*, since electrons in this band can contribute to charge conduction. For conductors the valence band and the conduction band are either the same or overlap in energy. For insulators and semiconductors the two bands are separated.

7.4 THERMAL EXCITATION   Most materials of practical interest are not in their "ground state" but have some excess energy or thermal excitation. This excitation energy plays an important part in electrical conduction properties, and consequently we will need some way to characterize and determine thermal effects on energy.

Consider a conductor in its lowest possible state. When there is no thermal energy (i.e. $T=0$), the probability of occupation of the various levels is unity up to a maximum energy, then drops abruptly to zero. If this occupancy distribution is described by the function $\omega(\epsilon)$ where ϵ is the energy above the lowest energy in the band, then the above description is equivalent to the graph shown in Fig. 7-9.

The function $\omega(\epsilon)$, as our picture showed, is 1 for all energies where the levels are occupied, and 0 for all energies where the levels are unoccupied. The energy where ω drops from one to zero (measured from the bottom of the band) is called the *Fermi energy*, E_F. In this situation† E_F is at an absolute energy of $-\phi$ (the work function). If the material has any thermal energy some of the electrons must be excited from their original levels into previously unoccupied levels. This excitation is more likely for electrons with lower energies (since more excitation energy is required to raise them to an unoccupied level). Correspondingly, the likelihood of occupying a state at increasingly greater energy decreases rapidly.

If these electrons did not have to obey the Pauli principle, their energy distribution would be described by the classical Maxwell-Boltzmann distribution. However, they do have to, and the distribution that accounts for the effect of the Pauli principle is called the Fermi-Dirac distribution (Fig. 7-10), and is given by

$$\omega(\epsilon) = \frac{1}{[e^{(\epsilon - E_F)/kT}] + 1}$$

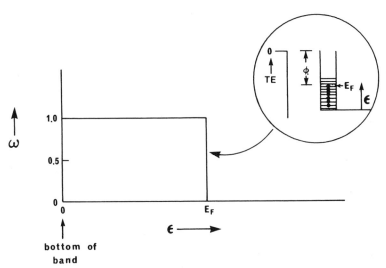

FIGURE 7-9 Graph of electron occupancy probability, ω, as a function of the energy from the bottom of the band, ϵ, for a material with $kT = 0$.

† In general the Fermi energy has a slight dependence on temperature, which we will ignore.

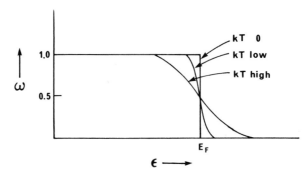

FIGURE 7-10 The Fermi-Dirac distribution for various values of kT.

The Fermi energy in this situation takes on the meaning that E_F is that energy where there is a 50 percent chance of occupancy. The number of vacancies created below E_F are equal in number and energy distribution about E_F to the occupancies that they generate above E_F. They are due to thermal excitation, and allow the existence of translation kinetic energy, as would be expected in an electron gas. As the material becomes hotter (kT increases) the slope of the curve decreases near E_F, and there is a greater probability of higher energy states being occupied. It is interesting to note (particularly in view of arguments made for thermionic emission) that for energies ϵ much greater than E_F and kT the function $\omega(\epsilon)$ can be approximated by

$$\omega(\epsilon \gg E_F, kT) = \frac{1}{e^{\epsilon/kT} + 1} = e^{-\epsilon/kT}$$

which is the same as the classical Maxwell-Boltmann distribution at high energies. From the general form of $\omega(\epsilon)$ it is clear that the conduction properties of materials are likely to have a strong temperature dependence.

This same treatment can be used to discuss the behavior of semiconductors. Here, however, it is necessary to exercise some care in deciding where to place the Fermi energy. One might at first guess that E_F for a semiconductor once again should be at the top of the conduction band, but this is not so. To see where it should be placed, consider what $\omega(\epsilon)$ would look like if a single electron were excited from E_v, the top of the (full) valence band to E_c, the bottom of the (empty) conduction band (Fig. 7-11).

From the mathematical symmetry of the Fermi-Dirac distribution about the energy E_F and the necessary symmetry of $\omega(\epsilon)$ between the top of the valence band and the bottom of the conduction band it is clear that E_F should be placed at the center of the energy gap. More detailed considerations* show that to a good approximation this is true in most cases, and we will take it as such.

* See Further Reading, Nanavati, sec. 2-10.

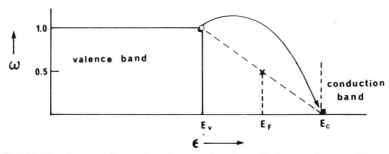

FIGURE 7-11 Fermi energy for semiconductor or insulator. E_v denotes the top of the valence band, and E_c denotes the bottom of the conduction band. One electron has been moved from the valence to the conduction band.

What are the relative probabilities of being able to thermally excite electrons into the conduction band for typical insulators and semiconductors at room temperature? We can estimate this by evaluating $\omega(\epsilon)$ for $\epsilon = E_c$. Although in principle the total probability of being in the conduction band should be obtained by integrating $\omega(\epsilon)$ for all energies $\epsilon > E_c$, it will suffice just to use $\omega(E_c)$, as this is the energy most likely to have an occupied state. Diamond has a gap energy E_g of approximately 5.5 eV. Therefore at room temperature ($kT = 0.026$ eV) the occupation probability for the bottom of the conduction band is

$$\omega(E_c) = \frac{1}{e^{E_c - (E_c - E_g/2)/kT} + 1}$$

$$= \frac{1}{e^{E_g/2kT} + 1}$$

$$= \frac{1}{e^{5.3/.052} + 1} = e^{-5.5/.052} \cong e^{-100} \cong 5 \times 10^{-45}$$

Less than one in every 10^{44} electrons in the valence band can have enough energy to be excited to the conduction band at room temperature. As there are only $\sim 10^{24}$ atoms in a gram-atomic weight of material it is not surprising that diamond is a good insulator! (Because we haven't calculated the actual density of states in the valence band of diamond, nor will we for any other material, we are unable to make any proper quantitative predictions for such things as conductivity; however we can see the basic causes of various properties and effects without such detail.)

For a typical semiconductor, say Ge ($E_g = 0.67$ eV), the probability is given by

$$\omega(E_c) = \frac{1}{e^{0.67/0.052} + 1} \cong e^{-0.67/0.052} = e^{-12.8}$$

$$= 2 \times 10^{-6}$$

Here ≈ 1 valence electron in a million can be excited to the bottom of the conduction band. It is certainly possible to find electrons in the conduction band for this material.

The chances are much smaller than for a conductor (where ω is of the order unity in the conduction band), but a reasonable number of electrons can still exist in the conduction band where they can contribute to electrical conduction properties. Notice the very strong dependence of ω for semiconductors on temperature. If we were to increase the temperature by 10 K, or 3 percent, at room temperature, the occupation probability would increase by approximately 30 percent. This sensitivity to temperature will increase as the gap energy of the semiconductor decreases.

7.5 CHARGE TRANSPORT BY HOLES Notice that in exciting an electron to the conduction band an unoccupied state in the valence band, or "hole", has been left behind. Our previously filled band is now only partly filled, and it is therefore capable of having some excitation energy, although only a very small number of electrons can be excited. This excitation process is most easily thought of as hole excitation, where the "hole" acts as if it were a positively charged body, which can contribute to the electrical conduction process. The electron excited to the conduction band leaves behind an "uncovered" net positive charge. All actual positive charge is fixed, of course, in the material, but an uncovered positive charge can be neutralized by an electron from a neighboring atom. Because the electrons can move, the hole can move about more or less freely (Fig. 7-12).

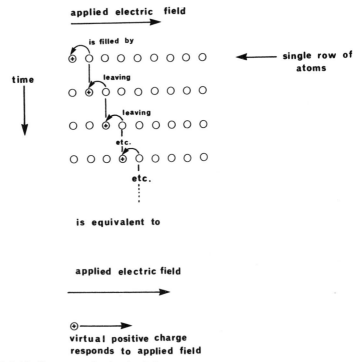

FIGURE 7-12 Charge conduction by holes.

The holes do not respond to a force, say an applied electric field, in the same way a free electron would, and so an *effective mass, m**, is assigned to them to account for the influence of other atoms on their mobility. Generally the effective mass of a hole is slightly greater than that of a conduction electron, and both are greater than that of a completely free charge.

7.6 SEMICONDUCTOR MATERIALS   From the previous considerations we expect semiconductor properties to exist in the diamond structured materials, such as the group IVA elements in the atomic tables. The energy gap in these materials decreases with increasing Z, as we might expect of the systematics of outer orbital quantum levels with increasing major quantum number n. However these materials are not the only ones that can form the tetrahedral structure characteristic of a band gap at valence-four. In fact the chemistry of such combinations of material as Ga-As, which have respectively 3 and 5 electrons in the same p shells instead of 4 as for Ge, can also lead to a tetrahedral structure and a semiconductor band behavior. This is also true for Al-P, Ga-P, and a large number of similar combinations, with bad gaps that can be varied more or less by "design." In fact an ever-increasing range of semiconductor materials are being generated to match specific demands for specialized uses, some of which will be discussed later.

7.7 DOPING   Not only is it possible to alter the gap energy of a semiconductor "by design," but it is also possible to change the other important parameter of semiconductor materials, the Fermi energy E_F, by the controlled introduction of "impurities" or dopants. From the calculation of the conduction band occupation probability we saw that only a few atoms in a million (depending on temperature and material)

Table 7-1   Values of Energy Gaps in Semiconductors

Crystal	E_g (eV)	Crystal	E_g (eV)
diamond	5.47–7.02	Si	1.12
BeSe	3.61	Ca_2Sn	0.9
AgI	2.82	GaSb	0.73
ZnSe	2.67	Ge	0.67–0.81
Cd S	2.41	Ca_2Pb	0.46
GaP	2.22–2.78	Pb S	0.41
Cu_2O	2.02	InAs	0.36
Ca_2Si	1.9	PbTe	0.31
CdSe	1.71	PbSe	0.27
AlSb	1.62–2.22	InSb	0.17
CdTe	1.52	Mg_2Sn	0.14–0.18
GaAs	1.42	Sn(grey)	0.08
InP	1.35	Ag_2Te	0.06

Source: Compiled from the data of Strehlow and Cook.

completely control the electrical properties of the semiconductor. Very clearly this means that semiconductor material used for devices must be extremely pure (to probably two orders of magnitude lower than the fraction causing the conduction, i.e., to something like one part in 10^8). However it *also* means that the introduction of small amounts of materials with known properties in a controlled fashion can be used to change the conduction characteristics of the semiconductor to whatever value may be desired. This process is called doping.

To see how this occurs consider the case of Si, illustrated in Fig. 7-13, where one of the Si atoms has been replaced by, for instance, a phosphorus atom, which has one electron more than Si.

Of the P valence electrons four will be used up in bonding the P atom into the lattice. Because the tetrahedral structure is favored, the fifth electron will find itself very weakly bound to the P atom, although it will still be bound since there is a net positive charge at the P site. We can estimate the magnitude of this binding by using the Bohr model, and one assumption that can be justified later. If the P atom were isolated its energy levels would be given by

$$E_B = \frac{E_0(Z_{eff})^2}{n^2} = \frac{E_0}{n^2}$$

and we would expect that $Z_{eff} = 1$, that is, complete shielding. However the P atom is *not* isolated; it is embedded in a matrix of Si atoms which have electrons around them and around the P atom. This electronic "sea" will certainly reduce the binding of the fifth electron in the P atom. This effect can be accounted for by remembering that

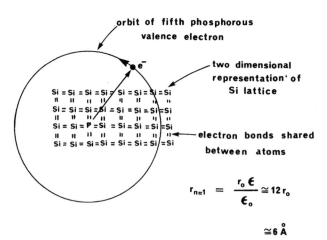

FIGURE 7-13 Orbit of least-bound phosphorus valence electron when embedded in a silicon lattice.

$$E_0 = \frac{m_e \, e^4 \, k^2}{2\hbar^2} \quad \text{where} \quad k = \frac{1}{4\pi\epsilon_0}$$

and ϵ_0 is the permittivity of free space (no charge present). We will therefore replace ϵ_0 with the permittivity for Si, which is $12\epsilon_0$. In this case, the lowest possible Bohr energy for the outermost P electron becomes*

$$E_B = E_0 \left(\frac{\epsilon_0}{\epsilon_{Si}}\right)^2 = \frac{E_0}{(12)^2} = 0.09 \text{ eV}$$

The electron is indeed found experimentally to be very weakly bound, by roughly this amount. Furthermore the Bohr radius corresponding to this binding energy is many angstroms, justifying our use of the bulk dielectric constant of Si to fix ϵ.

This outermost electron is bound so weakly to the P atom that its binding energy is comparable to thermal energies at room temperature. Consequently a very large fraction of all P atoms put into the lattice will contribute an electron to the conduction band. This can be indicated in the band diagram as shown in Fig. 7-14.

These controlled "impurities" are called *donors*, since they give up electrons. The corresponding material is labeled *n*-type because it has additional negative charge available for conduction. The loosely bound electrons must be given an amount of energy E_d to be excited into the conduction band.

Biasing

The presence of dopant atoms will affect the Fermi energy because they contribute excess electrons to the conduction band, so that the ratio of conduction electrons to valence holes is altered. (The donors do not produce free holes.) Consequently E_F shifts up towards the donor energy so that the form of the distribution function $\omega(\epsilon)$ will still properly describe the occupation distribution. Qualitatively E_F will depend

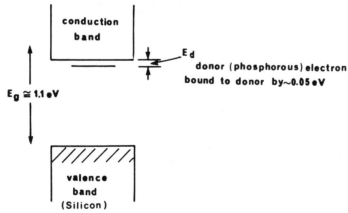

FIGURE 7-14 Energy of weakly bound donor electron relative to conduction band in silicon.

* Notice that we have let $n = 1$, since there are no other electrons in this new set of energy levels.

now on the concentration of the dopant material, the binding energy of the impurity levels, and the temperature.* At very low temperatures (Fig. 7-15) virtually all of the conduction due to the excitation of Si electrons will be "shut off" due to the magnitude of E_g compared to kT, but since $E_d \ll E_g$, the donor electrons can still be excited thermally to the conduction band, so that E_F approach value approximately $E_d/2$ below the conduction band. As the temperature increases, the excitation of the many more Si electrons can once again dominate the conduction behavior, and E_F will return to $E_{g/2}$.

There are several materials that can provide donor levels. The most obvious are As and Sb, although others can also be used, and provide a range of donor level energies.

Everything that has been said about donor materials—loosely bound electrons–can also be applied to materials which *lack* one electron to make up the valence-four arrangement for fitting a tetrahedral lattice. These dopants, such as Ga, act as acceptors of electrons, and create vacancies, or holes, in the valence band which can contribute to conduction. Their effect on band properties is illustrated in Fig. 7-16. This type of impurity is called an *acceptor* and produces positive charge conduction, or p-type semiconductor material. Electrons in the valence band require an amount of energy E_a to be excited into these acceptor levels.

The "binding energy" of these levels and their effect on E_F are analogous to those of n-type materials, except that the Fermi level is moved down as temperature decreases. When no dopant is present the semiconductor material is described as *intrinsic*.

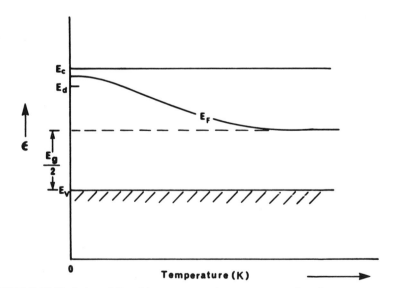

FIGURE 7-15 Variation of E_F with temperature in an n-type semiconductor.

* For a more detailed discussion of the calculation of E_F see Further Reading, Navon, p. 130.

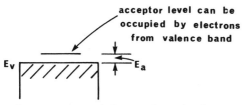

FIGURE 7-16 Energy of acceptor state relative to valence band.

7.8 SUPERCONDUCTORS When electrons are in the conduction band, either in conductors or semiconductors, the actual resistance to the flow of charge is determined by how frequently the electrons collide with the atomic lattice and lose their kinetic energy. It is this loss that makes it necessary to continuously do work on them whenever a current is desired, and gives materials their characteristic resistance. If there were no kinetic energy lost there would be no such thing as resistance; charges once set in motion would continue moving indefinitely. In fact there is a class of materials, called superconductors, where such behavior has been observed at very low temperatures (T $\lesssim$ 20 K). Several such elements, together with the critical temperature T_c below which they are superconducting, are listed in Table 7-3.

What property do these materials have that can inhibit electron energy loss, and why should it occur only at such low temperatures? To answer these questions we will have to consider how moving electrons interact with the lattice atoms in these materials.

Table 7-2 Ionization Energy for Donor and Acceptor States

| Doping Element | Silicon | | Germanium | |
	Donor E_c-E_d*	Acceptor E_a-E_v*	Donor E_c-E_d*	Acceptor E_a-E_v*
Li	0.033		0.0095	
Sb	0.039		0.0096	
P	0.044		0.012	
As	0.049		0.013	
B		0.045		0.010
Al		0.057		0.010
Ga		0.065		0.011
In		0.16		0.011

* in eV

Table 7-3 Critical Temperatures for Superconducting Elements

Z	Element	$T_c(K)$	Z	Element	$T_c(K)$	Z	Element	$T_c(K)$
13	Al	1.2	42	Mo	0.9	73	Ta	4.5
22	Ti	0.4	44	Ru	0.5	74	W	0.01
23	V	5.3	45	Rh	1.7	76	Os	0.7
30	Zn	0.9	48	Cd	0.5	77	Ir	0.1
40	Zr	0.8	50	Sn	3.7	80	Hg	4.2
41	Nb	9.3	57	La	~4.8	82	Pb	7.2

When an electron passes through the lattice it induces a polarization of the surrounding atoms and a slight distortion of the lattice structure, due to the Coulomb attraction between the polarized atoms and the electron itself. If the lattice is sufficiently "soft" to have a significant distortion which persists for a short time after the electron has passed, there is a *positive* charge concentration left behind momentarily. As illustrated in Fig. 7-17, this attracts a second electron to the site, most effectively one traveling in the opposite direction. These two electrons can be thought of as having undergone an *attractive* interaction through the medium of the lattice; they form a weakly bound pair state, called a *Cooper pair*.

This interaction can only take place for electrons near the Fermi energy, since they both must be moving, and the result of the binding between them is that a small gap is produced in the level occupancy curve, centered at E_F (Fig. 7-18). The interaction energy Δ is very small, on the order of 10^{-4} eV and the gap produced is comparably small (equal to 2Δ, since two electrons become paired). If the effect of this pairing interaction is to be significant, the temperature of the material must be very low, that is,

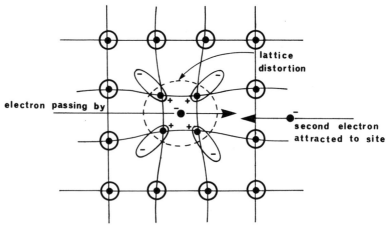

FIGURE 7-17 Attractive interaction between two electrons via lattice distortion in a superconducting material.

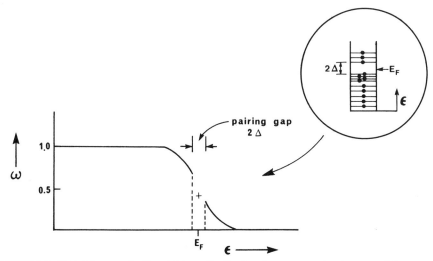

FIGURE 7-18 Pairing gap in Fermi-Dirac distribution for superconducting material.

$$300 \, K \times \frac{10^{-4} \, eV}{2.5 \times 10^{-2} \, eV} \cong 1 \, K$$

At these low temperatures in materials where the pairing interaction occurs, the paired electrons are unable to interact with the lattice and transfer energy to it exciting both electrons across the gap unless enough thermal energy is available to decouple the pair state. In this way momentum can be imparted to the pair (by an external electric field, for example), which cannot be dissipated by the normal resistive process, so long as there is not enough thermal energy to break up the pair. Currents, once set in motion, will not stop until the superconducting state is destroyed.

The complete absence of resistance in superconductors and consequent power savings should make them ideal in applications where large currents are required, for example, for magnetic field generation or power transmission. An example of one proposed use, a magnetic levitation vehicle, is shown in Fig. 7-19. In this vehicle both the levitating magnets and the magnets used for the vehicle's drive motor are superconducting.

However the necessity of maintaining the superconductors at the very low temperatures required with presently available superconducting material so far has proved to be a severe hindrance to the widespread application of superconductivity. This will change as higher-temperature superconductors and better refrigeration techniques are developed.

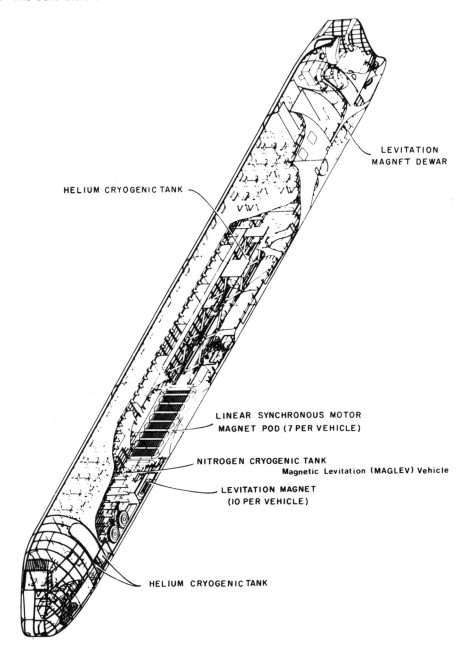

LEVITATION
MAGNET DEWAR

HELIUM CRYOGENIC TANK

LINEAR SYNCHRONOUS MOTOR
MAGNET POD (7 PER VEHICLE)

NITROGEN CRYOGENIC TANK
Magnetic Levitation (MAGLEV) Vehicle

LEVITATION MAGNET
(IO PER VEHICLE)

HELIUM CRYOGENIC TANK

FIGURE 7-19 Proposed design for magnetically levitated high speed vehicle. Both levitation and induction motor magnets are superconducting (courtesy of the Division of Mechanical Engineering, National Research Council, Ottawa, Canada).

FURTHER READING J. S. Blakemore, *Solid State Physics, 2nd ed.*, W. B. Saunders, 1974.

R. J. Elliot and A. F. Gibson, *An Introduction to Solid State Physics and its Applications*, Macmillan, 1974.

R. P. Nanavati, *Semiconductor Devices*, Intext, 1975.

M. A. Omar, *Elementary Solid State Physics*, Addison-Wesley, 1975.

B. B. Schwartz and S. Foner, Large-Scale Applications of Superconductivity, *Physics Today*, July 1977, p. 34.

PROBLEMS 1. Using the Bohr model, estimate what contact potential you would expect when potassium and cesium conductors were joined together. Their known work functions indicate that a contact potential of 1V should occur. Can you suggest the cause of the difference between this and your answer?

2. It was recognized that for electron microscopes the heat deposited in the sample by the beam could be significant. To get an idea of the magnitude of this problem, assume that the effect of the passage of an energetic electron is to deposit approximately the binding energy of one electron ($\sim$ 5 eV) in each atom it passes, and estimate what the general formula for the electron's energy loss per unit length should be. How does this estimate compare with what is observed.

3. What is the probability of an energy level 10 percent below E_F in a metal being occupied at (a) 300 K, (b) 500 K, (c) 70 K?

4. Show that the Fermi-Dirac distribution is symmetric about E_F.

5. Use the notion of energy bands to explain the following properties:

   (a) all metals are opaque to light of all wavelengths;

   (b) semiconductors are opaque to visible light but transparent to infrared;

   (c) many insulators are transparent to visible light.

6. Estimate the fractional change in conductivity (the change in probability of an electron being excited into the conduction band) for a 1°C temperature change at room temperature (300 K) for Si. How should this temperature-sensitivity vary with gap energy in general? What should the relative conductivity of Ge be compared to Si be at room temperature (using the above approximation)?

7. A small, but systematic variation of E_g with semiconductor temperature is observed. What do you expect this is due to, and should there be an increase or decrease of E_g with increasing temperature?

8. A small number of carbon atoms are put into a transparent crystal of quartz. They become "dissolved" in quartz due to the presence of the host material electrons and form triply charged ions. Using the Bohr model, estimate what frequency light these atoms will produce. Account for the negative charge from the host material by using the permittivity, given by

$$\frac{\epsilon}{\epsilon_0} = 4.3$$

9. How would you expect the effective mass of conduction electrons to compare with that of holes? Why?

10. Estimate what the value of E_d should be for As dopant in Ge using the Bohr model.

11. If the dopant in problem 10 were added to the extent of one atom per 10^4, what would be the position of the Fermi energy relative to the bottom of the conduction band at 300 K? (Think of the Fermi energy as the average energy from which *all* electrons in the conduction band are excited.)

12. The effective mass of a conduction electron is taken to be 1.2 m_e. What would the Bohr estimate for E_d be for As in Ge in this case?

13. A conduction electron that is bound to an As donor atom in Si de-excites by recombining with a hole. What frequency of light is emitted? Do you expect Si to be opaque or transparent to this frequency? Do you think it would be possible to determine dopant or impurity types by observing the properties of recombination light or perhaps by general optical characteristics?

14. Why should the correlation between electrons that form a Cooper pair be most likely if they are traveling in opposite directions?

15. The possibility of currents in a superconductor at very low temperatures seems inconsistent with the very close similarity in band structure of superconductor and semiconductor, and the known conduction properties of the latter. Explain this.

16. Is a Cooper pair a fermion or a boson? In what way do you think superconductivity might be altered if the Cooper pair were the other kind of particle?

EIGHT

APPLIED SOLID STATE PHYSICS

8.1 SEMICONDUCTOR JUNCTION DEVICES   What happens when two doped semiconductors—one *n*-type and one *p*-type—are placed in very close (atom-to-atom) contact?

Will the position of the conduction and valence bands change, or will they remain at the same value? Classically we would view this as bringing together two "gases" of electrons, each of which has a different "temperature," or mean kinetic energy—that is, different Fermi energies. The consequence of this would be that the hot gas and the cool gas would come to thermal equilibrium—the two Fermi energies would become equal. In fact this is what happens. When the two materials are brought together as illustrated in Fig. 8-1, the free electrons in the *n*-type material diffuse into the *p* region; they find lower energy states there. This net flow of electrons leaves behind a net unbalanced (and fixed), positive charge on the *n* side of the interface so that a potential difference is built up in the interface region between the *n* and *p* sides. This means that electrons now attempting to cross the interface to the right must overcome a potential barrier to reach the other side. Effectively the diffusion of electrons from the *n* to *p* side causes a depression of the total energy of conduction electrons on the *n* side and an increase of the total energy on the *p* side. This process continues until the Fermi energies are adjusted to be the same on both sides of the junction, as anticipated by the classical picture. The readjustment of the Fermi energies to be the same value in both materials ensures that electrons in the conduction band on both sides (and holes in the valence band) have a balanced distribution of energy, so that no further net flow of charge takes place. The two materials are at equilibrium. It is clear however that there is an internal potential difference built up between the two materials. This equilibration process, which occurs in metals as well as semiconductors, is what gives rise to a *contact potential* between different materials. The region in the immediate neighborhood of the junction, where the potential difference

161

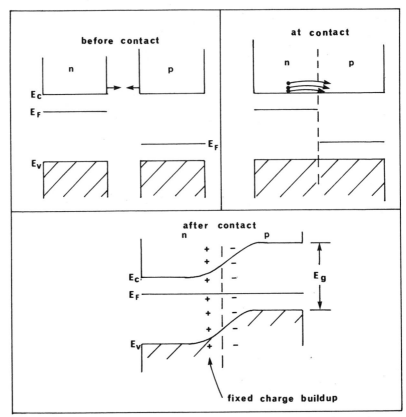

FIGURE 8-1 Development of level structure of *n-p* semiconductor junction as the two sides are brought together.

is generated, is called the *depletion layer*. When the junction is in equilibrium the internal electric field set up in the depletion layer due to the charge redistribution is accompanied by a removal of conduction electrons and holes from the depletion layer. The region is effectively depleted of charge carriers.

The *p-n* semiconductor junction we have just produced provides the basis for a great many semiconductor electronic devices. It is therefore of interest to look into the charge flow characteristics of this junction under various conditions. From the fact that there is a built-in potential difference between the two sides of the junction, one might think that an internal current must flow without any external bias. However this is not so, as we can see by estimating the net flow of electrons from one side of the junction to the other.

Because the Fermi energy on the *n*-side is close to the conduction band, there will be many more electrons on this side of the junction available to carry current than on the *p* side. However only those electrons (we will ignore the presence of holes for the time being) which have enough energy to surmount the potential barrier formed at the junction will be able to contribute to a net electron flow (i.e., diffusion of charge) dQ/dt to the right side (p region). Thus

$$\left(\frac{dQ}{dt}\right)_{\text{right}} \propto e^{-(E_g - E_F)/kT}$$

On the other hand, any electron which is excited to the conduction band on the p side will be able to flow to the left, since there is no potential barrier to stop them. This flow is proportional to

$$\left(\frac{dQ}{dt}\right)_{\text{left}} \propto e^{-(E_g - E_F)/kT}$$

which is just the probability of excitation to the conduction band. The net flow of charge across the junction will be the sum of these two currents

$$i = \text{const} \times \left[\left(\frac{dQ}{dt}\right)_{\text{right}} - \left(\frac{dQ}{dt}\right)_{\text{left}}\right]$$
$$= \text{const} \times \left(e^{-(E_g - E_F)/kT} - e^{-(E_g - E_F)/kT}\right) = 0$$

There is no net flow for this condition. This shouldn't be surprising since it is simply a confirmation of the statement that the p and n regions come to equilibrium when the Fermi energy reaches the same value in the two regions!

Now let us look at the current when an external bias is applied to the junction (Fig. 8-2).

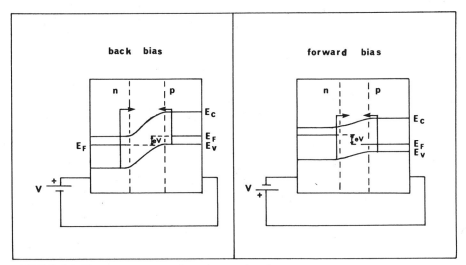

FIGURE 8-2 Electron movement through a p-n junction.

The effect of this external bias is to depress further the whole set of energy levels on the n side, including the Fermi energy.* A difference now exists between the Fermi energies on the two sides equal to eV, where V is the voltage of the battery. Now we will recalculate the net current through the device under these conditions. The flow to the right

$$\left(\frac{dQ}{dt}\right)_{\text{right}} \propto e^{-(\overbrace{E_g - E_F}^{E_0} + eV)/kT}$$

is decreased by the additional barrier height.

The flow to the left, however, is unchanged; it is still simply proportional to the probability of excitation to the conduction band.

$$\left(\frac{dQ}{dt}\right)_{\text{left}} \propto e^{-E_0/kT}$$

The net current is

$$i = const \times \left[e^{-(E_0 + eV)/kT} - e^{-E_0/kT}\right]$$

$$= const \times e^{-E_0/kT}\left(1 - e^{-eV/kT}\right)$$

For $V = 0$ the current is zero, as it must be, and for large values of V (compared to kT which is 0.025 eV) the current reaches a limiting value denoted by i_{leakage};

$$i_{\text{leakage}} = const \times e^{-E_0/kT}.$$

The limiting current is normally very small, and is called a leakage current. This bias condition is called a "back bias," since it effectively shuts off the current through the device.

Let us evaluate the current flow when the bias is applied in the opposite direction. The flow to the right is increased by the reduced barrier height

$$\left(\frac{dQ}{dt}\right)_{\text{right}} \propto e^{-(E_0 - eV)/kT}$$

whereas the flow to the left remains unchanged

$$\left(\frac{dQ}{dt}\right)_{\text{left}} \propto e^{-E_0/kT}$$

Under these circumstances the net current is

$$i = const \times \left[e^{-(E_0 - eV)/kT} - e^{-E_0/kT}\right]$$

$$= const \times e^{-E_0/kT}\left(e^{eV/kT} - 1\right)$$

* With no loss of generality we have assumed that the energy levels on the p-side remain fixed.

which is an exponentially *increasing* function of bias. The current in this bias condition (forward bias) is essentially unlimited.

The general characteristics of this junction can be summed up in the current versus voltage curve shown in Fig. 8-3.

Under a forward bias there is an exponential increase of current with voltage; large currents can pass through the junction. However when it is back-biased only a very small current flows, which is roughly independent of bias (since $e^{-eV/kT} \to 0$ *for eV* $>> kT$). In fact the junction acts as a unidirectional conductor; a diode, or rectifier.

So far we have ignored the holes. Their behavior is governed in the same way as the electrons, and does not qualitatively change our result; they just increase the magnitude of the current. Notwithstanding this, when detailed properties of individual semiconductors are examined the quantitative difference between electron and hole behavior we have just swept aside can take on considerable practical importance!

The operation of transistors can be understood readily in terms of what we have already developed for the diode. In fact a transistor is simply two diodes back-to-back, sharing a common element. The normal biasing arrangement shown in Fig. 8-4 is such that one diode is forward-biased and the second one is back-biased. The forward-biased diode provides a large current (the *n* section of this diode is called the emitter). The magnitude of the current is controlled by the value of the forward bias. The geometry of the common element is arranged so that as the electrons (in the configuration we have chosen) drift into the *p*-section (called the base), they are very unlikely to be collected in this part of the circuit. Most likely they will diffuse into the part of the *p* region where there is a strong potential gradient so that they are swept into the *n* region of the back-biased diode (called the collector). This will happen regardless of the magnitude of the back bias applied. There are basically two reasons for the small current collected in the base compared to the collector. First, the charge

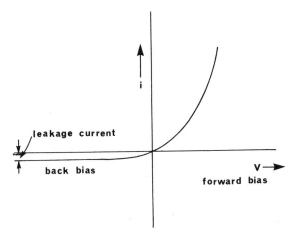

FIGURE 8-3 The *i-V* characteristics of a *p-n* junction.

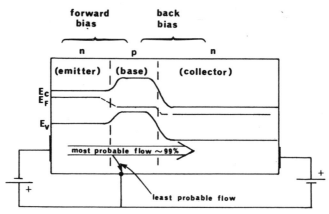

FIGURE 8-4 Electron movement through an *n-p-n* junction.

collection area is made much smaller for the base electrode than the base-collector interface area, so that the chance of reaching the base electrode is very small for the electrons. Second, the hole contribution to the emitter current is designed to be very small compared to the electron contribution. This can be accomplished by adequately high doping concentrations, and is necessary because holes can move easily into the emitter from the base.

It is not hard to see how the characteristics of transistors might be exploited. For instance a current variation in the emitter-base section can be generated by a small voltage variation, and can be made to pass through a large voltage drop in the base-collector section; small voltage signals can be amplified to large voltage signals. This is the basic property of transistors; many uses have been made of them and several modifications of the basic design are in common use.

8.2 ELECTRON TUNNELING   So far, the conduction characteristics we have looked at have dealt with the electron as essentially a "classical" particle in the conduction band. However we cannot always ignore the quantum properties of the electron and expect to understand everything that happens. One very important non-classical property is electron tunneling. Consider a diode that is strongly back-biased, so that electrons in the valence band of the *p* region are directly opposite (have the same energy as) unoccupied levels in the *n* region. Classically we would say that the electrons are confronted by a potential barrier (at the junction) which they must surmount (by thermal excitation) to get to the other side. However when we recognize that the electrons are actually described by wave functions, the question of getting past the barrier can also be asked in quantum-mechanical terms. This problem can be treated by a one-dimensional Schrödinger equation as a barrier penetration probability calculation.

The one-dimensional Schrödinger equation is

$$\frac{\partial^2 \psi}{\partial x^2} = \frac{-2m}{\hbar^2} [TE - V(x)]\, \psi \, .$$

On the right side of the barrier (Fig. 8-5), the solution of this equation gives the probability distribution for an electron with some (small) kinetic energy and zero potential energy. However there is also a solution to the Schrödinger equation *inside*

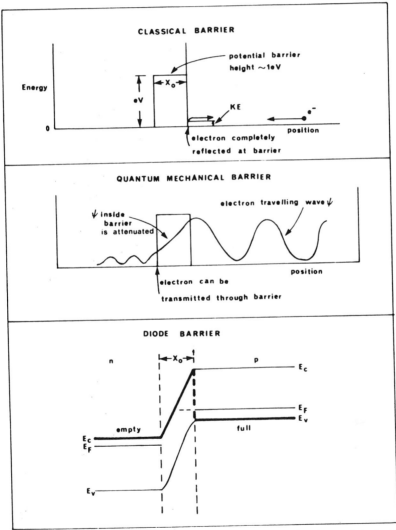

FIGURE 8-5 Barrier penetration by electron wave function and the shape of the barrier for tunneling in a back-biased diode.

the barrier region (which is classically forbidden) where the kinetic energy becomes less than the potential energy. In this region (assume $KE \ll 1\,\text{eV}$) the problem has the form

$$\frac{\partial^2 \psi}{\partial x^2} \cong + \frac{2m}{\hbar^2}\, eV\, \psi$$

which gives an exponential solution, where the main term is of the form

$$\psi_{(x)} = A\, e^{-\sqrt{2m\, eV/\hbar^2}\, x}$$

where A is the normalization constant, determined by the boundary conditions at the barrier edges.

The form of this solution shows that although the wave function extinguishes exponentially inside the barrier there is a finite probability of finding the electron on the other side of the barrier. If the barrier width is x_0 then the penetration probability is of the order $|\psi(x_0)|^2$, or

Penetration prob. $\propto e^{-2x_0/k}$

where

$$\frac{2}{k} = 2\sqrt{\frac{2m\, eV}{\hbar^2}}$$

and $k/2$ is the length in which the penetration probability falls off by $1/e$. For an electron tunneling through a 1 eV barrier this length is

$$\frac{k}{2} = \frac{1}{2}\sqrt{\frac{(1 \times 10^{-34})^2}{2 \times 10^{-30} \times 1.6 \times 10^{-19} \times 1}} = \frac{1}{2}\sqrt{\frac{10^{-68}}{3 \times 10^{-49}}}$$

$$\cong \frac{2}{2} \times 10^{-10}\ \text{m}$$

$$\cong 1\ \text{Å}$$

This calculation shows a characteristic tunneling length of the order of an angstrom; however, as can be seen in Fig. 8-5, the shape of the barrier encountered in the diode junction is approximately triangular, rather than square, and its effective width depends on the applied back bias. The penetration calculation for such a barrier* shows a strong dependence on the junction width and magnitude of the back bias, as for the square barrier, but more importantly it indicates that tunneling can occur for distances up to $\sim 10^2$ Å, much greater than our simple calculation suggested.

How can such abrupt changes in conduction band energies be produced? The answer is clear if we think back to the physical process that caused the change: the diffusion of electrons from the n side of the junctions (and holes on the p side) to the other

* See Further Reading, Ziman, sec. 6.8

side, "uncovering" fixed positive charges which depress the conduction band energy on the n side. The amount of charge required depends on the difference of the Fermi energies in the two types of semiconductor, but the distance back from the junction that will be depleted in order to uncover that charge depends on the density of conduction electrons available, and hence the density of dopants. The narrower the transition region required, the greater the doping necessary. In order to reduce the transition layer thickness (normally $\sim 10,000$ Å) to that necessary to allow tunneling, very heavy doping is required (on the order of 10^{20} dopant atoms/cm^3 $- \sim 10^3$ times that used for normal semiconductor material).

8.3 ZENER, AVALANCHE, AND TUNNEL DIODE   When a semiconductor junction possesses the properties necessary for tunneling, several useful effects can be obtained. One such effect can be observed in the reverse bias characteristic of a heavily doped diode. At some reverse bias, characteristically a few volts (enough to make the n conduction band overlap the opposite valence band in energy), tunneling will suddenly take place and the diode's reverse bias "resistance" disappears. Such diodes, called Zener diodes, can be used to control or to limit potential differences in circuits, as they are unable to sustain a back bias greater than their breakdown voltage, as illustrated in Fig. 8-6.

The process involved in the Zener breakdown should not be confused with another back-bias breakdown mechanism, called *avalanche breakdown*. This is a completely classical process in which the electrons involved in the reverse current are accelerated

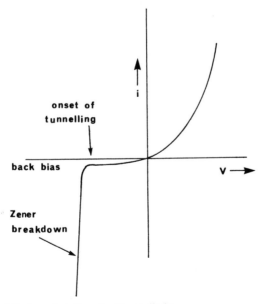

FIGURE 8-6 The i-V characteristics of a Zener diode.

to a sufficiently high kinetic energy, between collisions with the lattice while passing through the depletion region, that they are able to knock bound electrons out of the lattice and multiply the back-bias current. For this to happen the back bias must be rather higher than that for Zener breakdown, and breakdown voltages as high as 100 volts can be produced. Conditions characteristic of avalanche breakdown and Zener breakdown are shown in Fig. 8-7.

The forward bias characteristics of a diode also can be modified very usefully by the tunneling effect. In this case the device is called a *tunnel diode*, and is doped so heavily that the Fermi energies are actually pushed into the conduction band in the *n* region, and into the valence band on the *p* side (Fig. 8-8). This not only leads to a very narrow junction (~ 10 Å) but produces a major modification to the *i-V* characteristic curve for the diode.

The depletion layer is so narrow that tunneling occurs even with a very small bias. Consider the back bias condition; as soon as any bias is placed on the device, electrons on the *p* side are directly opposite (physically and energetically) a vacant region in the *n* side, so tunneling can take place. The diode acts as a normal conductor. Similarly for a small forward bias the device will act as a conductor, up to the point where the bottom of the conduction band on the *n* side is at the same energy as the top of the valence band on the *p* side. At this point tunneling is cut off, since there are no longer energetically allowed states opposite the conduction electrons. At forward biases greater than the tunneling cut-off value all conduction is carried out by thermal excitation, as in a normal diode. The overall *i-V* characteristics of this diode are a composite of tunneling and thermal properties (Fig. 8-9), and have a rather unusual shape. In the part of the curve where the current changes from tunneling to thermal conduction there is a decided *decrease* of current with increasing bias.

If the device is forward-biased so that the current is at the peak of the tunneling conduction curve, the diode can be used as a very fast voltage trigger. Even a very small current pulse passing through the diode in this condition causes an immediate change in the voltage across the diode from V_0 to V_1. The basic reason the voltage jump can be so fast (≈ 1ns switching time) is that the junction is extremely narrow.

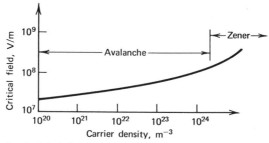

FIGURE 8-7 Graph of critical field for junction breakdown versus carrier density for *n* type silicon indicating both the avalanche and Zener regions. From David H. Navon, *Electronic Materials and Devices*, page 183. Copyright © 1975 by Houghton Mifflin Co. Reprinted by permission.

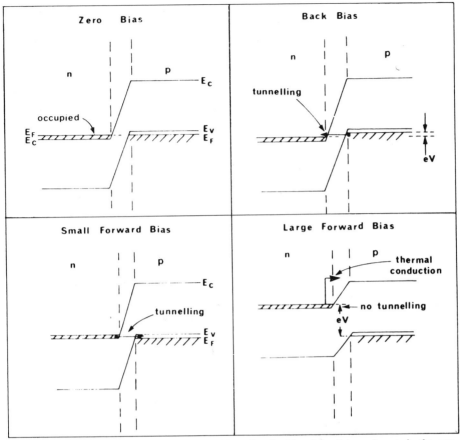

FIGURE 8-8 Operation of a tunnel diode. Note that the Fermi level is no longer in the gap, due to heavy doping.

8.4 NEGATIVE RESISTANCE AND THE GUNN EFFECT

Another kind of semiconductor application can be understood when we recognize in the above i vs V curve that the region of negative slope corresponds to a *negative resistance*.

This property can be used to drive an oscillator circuit. Normally a resonance circuit has a finite resistance so that any initially present oscillations decay away as the stored energy is dissipated in the resistance. However if a device with the above characteristics, biased into the negative resistance region, is added to the circuit with a dynamic negative resistance equal to the positive resistance already present, the oscillations could continue undamped, with the negative resistance device feeding in power to cancel that dissipated by the real resistance. This leaves the question of what semiconductor device has characteristics that are most effective for this kind of application and what physical processes give rise to those characteristics.

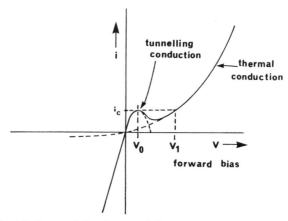

FIGURE 8-9 The *i-V* characteristic of a tunnel diode, showing initial current, i_c, where rapid switching from bias V_0 to V_1 can occur.

Although the tunnel diode has the necessary negative resistance it cannot normally operate in this region at very high power levels and therefore is not well suited to many oscillator applications. A widely used device for generating high-frequency oscillations is simply a piece of bulk semiconductor such as GaAs, which exhibits the *Gunn effect*.

Up to now we have not paid any attention to the electrons once they are in the conduction band. The Gunn effect arises from *subsidiary conduction bands*, a more detailed property of semiconductor materials than we have dealt with so far. Electrons traveling in the conduction bands of materials are not completely free, since they are continuously scattering from the lattice atoms. These moving electrons also have wave properties and the interaction of the electron waves and the periodic lattice structure can influence strongly the momenta and energies of the electrons (in fact a more complete and precise band theory than we have used can be constructed from such considerations). In GaAs, electrons can be excited up to the conduction band most readily ($E_g = 1.43$ eV) without a change in the direction of their wave vector **k**. (Such an excitation is said to involve a *direct gap*.) This process can be thought of by considering the valence electrons as a standing wave in one particular direction,* which is then excited up to the conduction band and becomes a traveling wave in that *same direction*. It is also possible (Fig. 8-10) with approximately 1.78 eV excitation energy to promote electrons from the valence band to a mode of motion through the lattice corresponding to a change in wave vector; that is, another allowed propagation direction through the lattice. This subsidiary conduction band (an *indirect gap*) has associated with it a greater effective mass than the lower-energy direct gap band, so that electrons excited to this subsidiary band respond more slowly to an applied electric field. (In silicon and germanium the indirect gap is at a lower excitation energy so the major excitation mechanism is indirect.)

* Remember that a standing wave is a superposition of two traveling waves moving in opposite directions.

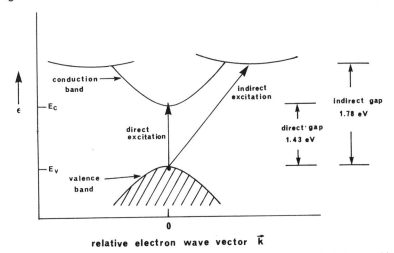

FIGURE 8-10 Band structure accounting for electron wave vector property. Direct and indirect gap energies shown are for GaAs.

The combination of direct and indirect conduction bands in GaAs leads to a characteristic variation of electron drift velocity, v_d (the average velocity attained between collisions with the lattice) with applied electric field. At low electric fields all conduction electrons are due to direct excitation, and have a relatively low effective mass and therefore can attain high drift velocities. When the electric field reaches a sufficiently high value a number of the electrons will be raised into the subsidiary band with the consequence that they drift at a lower velocity. As the field is increased further, greater numbers of electrons are raised into this band so that the average drift velocity now decreases with increasing electric field. Finally when virtually all electrons are in this subsidiary band, the overall drift velocity begins once again to increase with applied electric field due to the response of the "heavier" electrons to an increase in field. The resulting v_d vs electric field curve takes on the shape shown in Fig. 8-11.

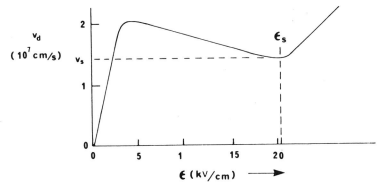

FIGURE 8-11 Drift velocity v_d versus applied electric field ϵ for GaAs.

However for a given piece of material, drift velocity is proportional to current, and the electric field is proportional to the applied voltage, so the above curve is equivalent to an i vs V plot, which shows that GaAs has a characteristic negative resistance region of operation.

With this information we can see how to use GaAs to make an oscillator. If a voltage is placed across a piece of GaAs to produce an electric field corresponding to the region of negative resistance, any small increase in electron density, present due to random fluctuations, will cause a local increase in the electric field (by Gauss' law) and will cause a reduction in the local value of drift velocity. This, in turn, will lead to an increase in local charge density since electrons previously behind the group will catch up with it and be slowed down due to the increased field. This process will continue increasing the field and charge density in the local region, creating a *space charge domain* and reducing the charge and (again by Gauss' law) electric field elsewhere until the drift velocity in the domain reaches its minimum high-field value v_s. At this point the domain reaches a stable state and traverses the length of the semiconductor and passes into the external circuit. When the charge pulse leaves the GaAs the electric field values return to their initial values and the whole process starts over again. In this way a train of charge pulses can be set up by supplying a DC bias to provide the necessary electric field. Characteristic frequencies can be as high as 10^{10} Hz for such Gunn effect oscillators.

8.5 OPTICAL PROPERTIES OF SEMICONDUCTORS   There is another interesting property of semiconductors, and that is how their electrical characteristics are influenced by the presence of light. When dealing with molecular gases a plot of opacity versus optical frequency was generated. This can be repeated for semiconductors and the result is shown in Fig. 8-12. For an intrinsic semiconductor we can

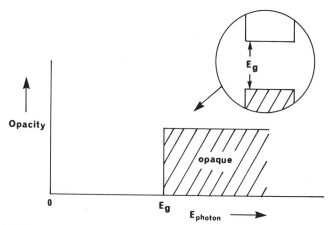

FIGURE 8-12 Optical properties of semiconductor material.

expect the material to be transparent to light of low frequencies ($E_{ph} < E_g$) and opaque for frequencies corresponding to energies greater than E_g. Recalling the energy range of visible photons (red light, $E_{ph} \cong 1.7$ eV; violet light, $E_{ph} \cong 3.1$ eV) and the range of gaps for semiconductors, we generally expect semiconductors to be transparent in the infrared region, and opaque in the visible. Of course the opacity of the semiconductor material can be modified by the presence of dopants, which will increase the opacity either at the low frequency end (p type) or at energies just below E_g (n type).

A basic electrical consequence of the optical opacity of semiconductor material is that when incident light is absorbed by a semiconductor, the number of electrons excited to the conduction band is increased so that the material's conductivity becomes greater. This increase can be quite marked, such as for CdS ($E_g = 2.4$ eV), where the conductivity can be increased a thousandfold by shining an intense light on it.

One of the more interesting uses of photoresistivity—the dependence of resistance on incident light intensity—is found in the vidicon. Imagine the following, rather sophisticated, technique for measuring resistance sketched in Fig. 8-13. If a capacitor C is charged to a potential V by the battery and then allowed to discharge through the resistor R for a definite amount of time t_0, then the resistance can be determined by

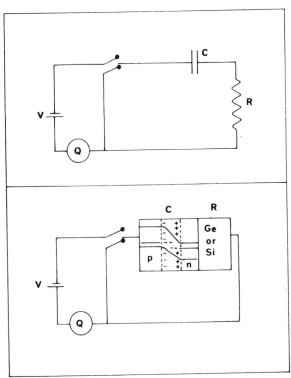

FIGURE 8-13 Current for measuring resistance by means of charge depletion of capacitor.

measuring the charge required to recharge the capacitor, since the amount lost is directly related

$$Q(t) = CV \, e^{-t_0 / RC}$$

to the value of R. In this way any changes in R can be sensed by a change in the charge required to recharge the capacitor. This device, miniaturized and constructed from semiconductor materials can be used to measure light intensities and variations, through photoresistivity. The p-n junction properties are such that when back-biased the conduction through the diode will be reduced to the very small leakage current. Consequently it can be thought of as a "leaky" capacitor which will store the injected charge for a finite time, which is determined by the magnitude of the leakage current.

The back-biased diode "capacitor" can be given a known charge by injecting electrons with a kinetic energy corresponding to the desired back bias on the diode. In this way charging ceases at a potential V equal to the electron beam accelerating potential, since at that value no more current can flow. (An electron beam is used because it can act as a "moveable wire.") If the diode is connected by an external circuit which includes a semiconductor photoresistor of the appropriate value, the discharge time constant of the external circuit can be several orders of magnitude shorter than that for the internal leakage path, so that the value of the external resistor can be determined in the way previously described. Since the value of the resistance is controlled by the flux of photons (with $h\nu \geq E_g$) on the photoresistor while the diode is discharging, this measurement also gives a determination of the light intensity on the photoresistor.

Using this technique a whole picture can be built up by means of a large array of closely spaced charge-storing diodes (Fig. 8-14). These are charged up by a sweeping electron beam, and the opposite side (the photoresistor) is illuminated by the focused image of some object. The elements of the diode array are discharged by an amount that is proportional to the light intensity at that particular point. The electron beam sweeps in an X-Y (raster) scan at a constant rate, and recharges the diodes to a voltage equal to the electron beam energy. The restoring charge required for each diode in the array can be measured and then used to electronically reproduce the image. Since some semiconducting materials have gap energies corresponding to infrared frequencies it is possible to take infrared "pictures" of objects, as is done in sniper scopes. There are many variations of this basic technique in use.*

8.6 SOLAR CELL   The increase in charge carriers due to light striking a semiconductor can be used in a somewhat different way if, instead of shining light on a piece of semiconductor to change its conductivity, the light is made to fall onto a diode as shown in Fig. 8-15.

The incident photon will excite electrons on the p-side up to the conduction band. If this electron diffuses into the region with the electric field it will be driven towards

* See Further Reading, Biberman and Nudelman, Ch. 13.

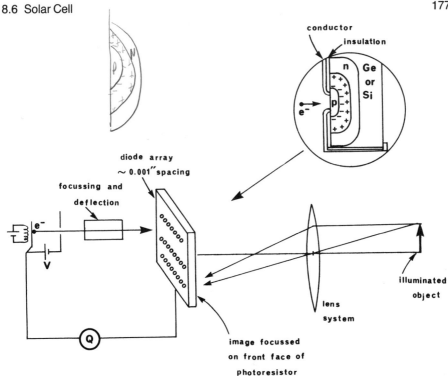

FIGURE 8-14 Arrangement of vidicon, showing charge storage diodes, photoresistor, and charging electron beam.

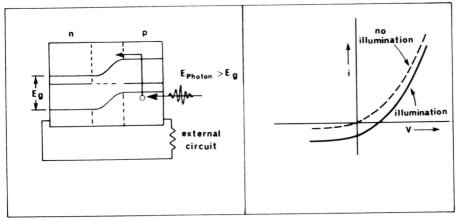

FIGURE 8-15 Action of photons on diode and corresponding i-V characteristic for illuminated diode.

the *n* side of the diode, by the electric field, and will contribute to the flow of electrons to the *n* side of the conductor. This will cause an imbalance of charge crossing the diode, and act as a current source. The magnitude of this current source is determined entirely by the light flux on the *p* side of the diode and not by bias conditions. The light energy acts as a "pump" for the solar cell, but will only work for frequencies greater than those corresponding to the semiconductor gap energy.

8.7 ELECTRON OR PHOTON DETECTOR   This same basic diode arrangement can be used to detect individual energetic photons or electrons (or any charged particle, for that matter). This can be done by placing a large back bias on the diode to increase the extent of the region where a potential gradient exists.

This field exists because of the displacement of the positive and negative charges required to adjust the Fermi energies. The extent of the region can be increased further by inserting a section of intrinsic semiconductor, so that most of the volume of the diode is in an electric field region.

If an energetic photon or electron strikes an atom in this intrinsic region it can give up some or all of its energy by exciting electrons from the valence band to the conduction band, creating electron-hole pairs as shown in Fig. 8-16. These electron-hole pairs are in the electric field, and so are driven by the field to the *n*- and *p*-regions of the detector. By "driving" the electron and hole through the potential difference, the field does work on the charge pair; this work, $W = q\Delta V$, can be considerably greater than the energy of the initial photon or electron. For example if a 2 eV photon releases an electron-hole pair in the intrinsic region, the diode (if it is back-biased at 4000 V) will do 4000 eV of work in removing the charge pair from

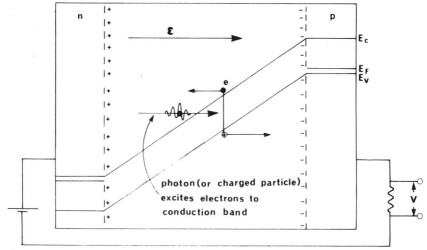

FIGURE 8-16 Operation of semiconductor radiation detector.

the intrinsic region. This work can be detected, and effectively constitutes a two-thousandfold amplification of the incident energy "signal." Such a device often is used in scanning electron microscopes to detect electrons, X rays, γ rays, or visible photons.

A major reason for the usefulness of such semiconductor radiation detectors is the fact that they are capable of determining the energy of radiation that produces a signal. By making the intrinsic region of the detector thick enough to absorb completely the energy of the incident charged particle or photon, all of the projectile's energy can be expended by exciting electrons from the valence band to the conduction band (plus giving them some excess kinetic energy). The number of conduction electrons produced, and therefore the magnitude of the charge driven through the diode, is directly proportional to the absorbed energy, and provides a straightforward measure of the energy deposited in the detector by the incident radiation.

The internal electric field in semiconductor radiation detectors is not only capable of generating signals that provide a measure of the radiation energy, but could also be used in principle to determine where within the detector the radiation arrived. One possible way to do this is illustrated in Fig. 8-17. Radiation striking the detector at some point in the intrinsic region liberates electrons and holes, which are driven in a straight line toward the two electrodes. If each of these electrodes is divided into a set of closely spaced parallel strips, the two sets of strips being mutually perpendicular, the x and y coordinates of the point where the electron-hole pairs are generated

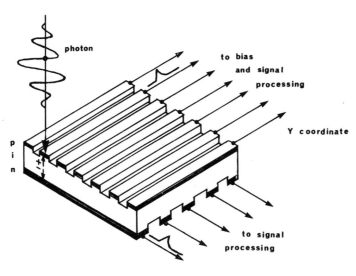

FIGURE 8-17 Semiconductor x-y-position-sensitive photon detector. The direction of the internal electric field ensures that induced charge is impressed on those electrodes that have the same x and y coordinates as the original interaction position.

is given by which strip carries the induced charge signal. Because the charge only travels along the field lines, only those strips directly connected to the electron-hole pairs by the electric field lines will produce any signal. Such a technique has been used to produce a γ-ray "camera" using a center-to-center strip spacing of 2 mm. A picture of a distributed source of 122 keV γ rays (in a rat) is shown in Fig. 8-18. As the basic detector area is only 6.4 × 3.2 cm², the full-length picture is made from a composite of equal-exposure sections.

8.8 ELECTRON-HOLE RECOMBINATION So far we have looked into the characteristics of semiconductor currents, and light generation of conduction electrons and holes. Quite clearly, however, it not only should be possible to create charge carriers by the absorption of light, but also to create light by the recombination of electrons and holes (Fig. 8-19).

The second process—recombination—is relatively infrequent in indirect gap semiconductors, since both electron and hole must exist at the same site long enough to de-excite. The probability of an individual conduction electron being "at" a particular position is small to start with ($\lesssim 10^{-3}$).* This must then be multiplied by the prob-

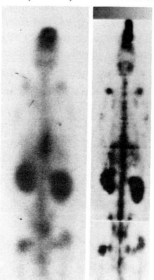

FIGURE 8-18 A picture of a rat taken with an x-y position-sensitive semiconductor "camera" (*right*). The rat has been injected with ^{99}Tc which emits 122 keV photons; the image is a "contact print" obtained by placing an array of very closely spaced collimators between the object and the detector. For comparison a high-resolution pinhole camera picture is shown on the left-hand side. Courtesy of L. Kaufman, Radiologic Imaging Laboratory, Department of Radiology, University of California, San Francisco.

* We are assuming that one atom in a thousand provides a conduction electron.

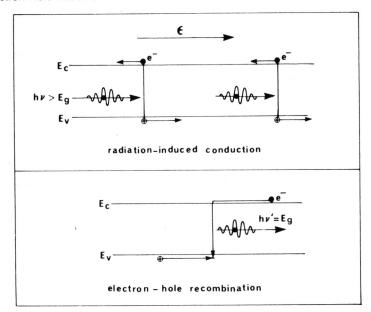

FIGURE 8-19 Production of radiation by electron-hole recombination.

ability of a hole existing there as well (a comparably small number) so that recombination processes are relatively improbable.

If we want to enhance the likelihood of this recombination process in order to produce light, it is clear that we must at the very least increase the number of conduction electrons and holes as much as possible. This can be done by heavy doping, as in the tunnel diode. The use of a tunnel diode guarantees a high concentration of both electrons and holes, but as Fig. 8-20 shows, the electrons and holes are *not* in the same place, which they must be to recombine. However this can be accomplished by applying a large forward bias to the diode. The combination of heavy doping and a forward bias comparable to the gap energy provides a high probability of having both conduction electrons and holes in the same place at the same time, and therefore a high recombination probability. This configuration gives considerable light output, and is known as a *L*ight *E*mitting *D*iode, or LED. There are several materials that are used for light emission in the visible region. Some of them and their gap energies are listed in Table 8-1.

The fact that the emitted photons have an energy comparable to the semiconductor gap energy immediately suggests a major practical difficulty for LEDs, that is, self-absorption of the generated light. This is a problem because it means that there will be a high index of refraction for the emitted frequency while in the semiconductor material. As an example, GaAsP has an index of refraction n ≅ 3.5, indicating that the effective velocity of the light is reduced by approximately a factor of three by repeated absorption and re-emission of the light. In practical terms this means that

TABLE 8-1   Gap Energies of Selected Semiconductors

	E_g(eV)	
AgI	2.8	
ZnSe	2.7	blue
CdS	2.4	green
GaP	2.3	
CdSe	1.7	red
GaAsP	~1.6	

there is a very small critical angle ($\theta_c \cong 17^0$ for GaAsP), so that only a small fraction of the generated light is emitted from the diode. This problem can be alleviated by coating the semiconductor with a material (or materials) with an intermediate index of refraction, thus increasing the critical angle and allowing more of the light to escape.

8.9 THE SOLID STATE LASER   The condition necessary for a system to lase (assuming a two-state system) is that the higher energy state of the system have a

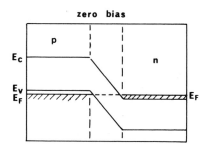

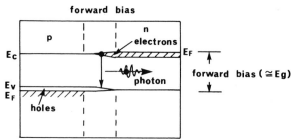

FIGURE 8-20 Light emission from a *p-n* diode.

greater population than the lower one. It is quite clear that such a situation can be created in a forward-biased tunnel diode, since a large population of electrons and holes (i.e., excited states) exist together. If this high concentration is maintained by driving the diode hard to drift as many electrons and holes into the recombination region as possible, lasing action is possible. The density of photons in the active region can be increased—and therefore the probability of stimulated emission—by grading the index of refraction* of the semiconductor, so as to increase the amount of total internal reflection (the opposite to what is done for LEDs). End mirrors for the laser ''cavity'' can be produced by cleaving the crystalline semiconductor material.

The concentration of electrons and holes can be enhanced further by using a so-called ''pinstripe geometry.'' Here, as shown in Fig. 8-21 the contact to the p-region of the diode is confined to a narrow stripe. As the p-region is relatively thin (a few microns) this effectively confines the holes in the p-region to the central part of the diode just under the stripe. This provides a further increase in concentration of holes (and therefore electrons), improving the performance of the laser.

The requirement of high current densities for maintaining a population inversion and a high photon density means that the diodes must withstand rather heavy heat loading†. Consequently the diodes must be adequately cooled, and pulsed operation, rather than continuous, is often a more practical mode of operation. Solid state lasers have been made to operate with optical powers up to 50 W and over a wavelength range from 0.32 μm to 16.5 μm.

8.10 CRYSTAL IMPERFECTIONS The desirability of making diodes with graded indices of refraction brings up an important property of real semiconductor materials which we have ignored thus far. We have assumed ideal crystalline prop-

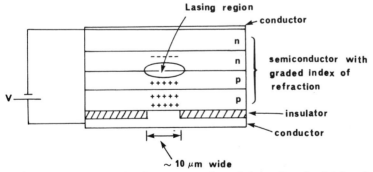

FIGURE 8-21 Pin-stripe laser showing concentration of charge for stimulated emission, and graded index of refraction to form resonant cavity. The laser beam is emitted out from the plane of the drawing.

* The index of refraction can be controlled by the choice of dopant density.

† See Further Reading, Navon, Sec. 11.3.

erties: no imperfections or defects such as vacancies (no atom where there should be one), interstitials (atoms in spaces in which they shouldn't be), dislocations (whole lines of atoms displaced from their proper places), and so on. To give an example of the real importance of such imperfections, it has been found in multilayer diode junctions for LEDs and lasers that the use of too large a change in lattice spacing ($\geqslant$ 1%) at the various interfaces can produce a large number of defects, and that these can "trap" electrons and holes and allow them to recombine without emitting photons of the desired frequency.

A way to understand how defects, such as vacancies, can be responsible for this behavior is indicated in Fig. 8-22. Obviously the presence of a vacancy in a lattice has a strong effect on the electron energies in the sites immediately adjacent to the vacancy and will also generally distort the lattice structure out to several lattice spacings. This distortion can be thought of as generating a number of atoms whose energy

Semiconductor   Imperfections

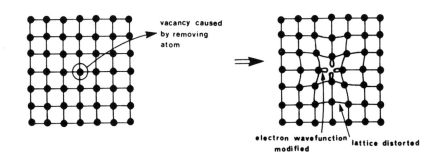

Effect On Band Properties

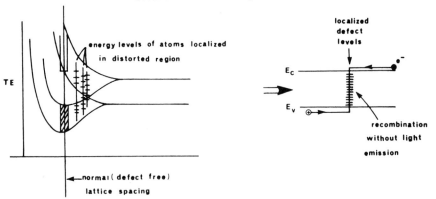

FIGURE 8-22 Lattice vacancy leading to distortions and localized levels in semiconductor gap.

levels (because of their modified nearest-neighbor spacing) do not fit into the previous band picture. They create localized energy states that can fall *within* the gap, and be quite complex in detail. The effect of the distortion, together with more complex vibrational modes of the lattice possible at the vacancy site leads to local "impurity" levels within the energy gap. These levels lead to strong localization, or trapping, of both electrons and holes, and enhance their rate of recombination. Due to the complexity of the local level structure, recombination can occur without the emission of the desired photon. This is illustrated strikingly in Fig. 8-23 where a mismatch between two semiconducting layers has produced regular arrays of dislocations that can be

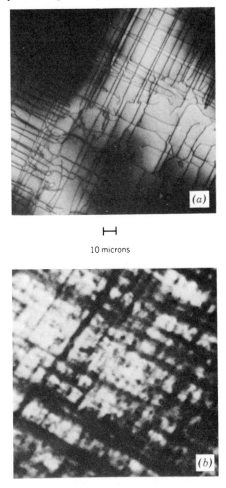

10 microns

FIGURE 8-23 Misfit dislocation in $In_x Ga_{1-x}$ P layer on a GaP substrate. A comparison of a transmission electron micrograph (*a*) and luminescence scan (*b*) shows that dislocations near the surface appear as nonradiative regions. This, and Fig. 8-24 are from "Light Sources" by H. Kressel, I. Ladany, M. Ettenberg, and H. Lockwood, *Physics Today,* May 1976.

seen in the electron microscope view of the material. The same pattern is observed in the lack of photon emission from electron-hole recombination, confirming the association of structural defects with radiationless recombination.

Although undesired diode characteristics can be introduced by uncontrolled modifications of the atom-atom spacing in semiconductor materials such as occur when vacancies are present, it is possible to modify the average interaction energy via the lattice parameter, in a controlled fashion. The use of multiple element alloys of continuously variable composition makes it possible to vary the lattice parameter over a fairly wide range of values, and consequently change the gap energy considerably. The use of quaternary alloys as shown in Fig. 8-24 provides an even greater versatility in range of lattice parameter, band gap, and index of refraction at any given photon energy.

8.11 PRESSURE SENSITIVITY   The close relation between lattice spacing and semiconductor energy gap is something that we expect from our basic picture of semiconducting materials, and is exploited in the quaternary alloy semiconductor. It also should be possible to make use of this relationship in a rather more direct way. By squeezing very hard on a material, such as a spring, we are able to compress it

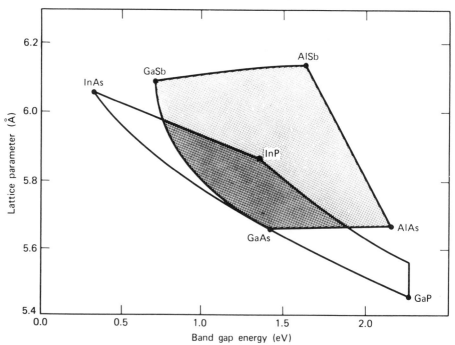

FIGURE 8-24 Extent of band gaps and lattice parameters covered by the quarternary alloys $Al_x Ga_{1-x} As_y Sb_{1-y}$ and $In_x Ga_{1-x} As_y P_{1-y}$.

so that the atom spacing is reduced from the normal equilibrium value. If we do this to a solid-state laser or LED we expect the same thing to happen, and as a result the band gap should change (our simple model would predict that it should increase). This changes the emission frequency of the material, and therefore provides an indication of the amount of pressure applied.

Because normal semiconductor materials like silicon and germanium are rather hard, it would take unusually high pressures to produce a noticeable effect and consequently this property would be of little practical use. However certain lead compounds called chalcogenides (compounds with group IV and group VI elements) from which semiconductor material can be formed ($E_g \simeq 0.1 \to .01$ eV) are much more compressible. In principle it is possible to construct an LED or a laser in the far infrared whose frequency could be used to measure directly the pressure on the device or which could use a varying pressure to frequency-modulate the laser. Such pressure modification has been used to tune PbSe lasers over a range from 8 μm to 22 μm.

FURTHER READING L. M. Biberman and S. Nudelman, eds., *Photoelectronic Imaging Devices*, vol. 2, Plenum, 1971.

C. A. Hogarth, ed., *Materials Used in Semiconductor Devices*, Wiley, 1965.

L. Kaufman, K. Hosier, V. Lorenz, D. Shea, J. Hoeninger, A. Ching, M. Okerlund, R. S. Hattner, D. C. Price, S. Williams, J. H. Ewins, G. A. Armantrout, D. C. Camp and K. Lee, Imaging Characteristics of a Small Germanium Camera, *Investigative Radiology* 13 (1978), p. 223.

J. H. Leck, *Theory of Semiconductor Junction Devices*, Pergamon, 1967.

T. S. Moss, *Optical Properties of Semiconductors*, Butterworth, 1961.

R. P. Nanavati, *Semiconductor Devices*, Intext, 1975.

D. H. Navon, *Electronic Materials and Devices*, Houghton Mifflin, 1975.

S. M. Sze, *Physics of Semiconductor Devices*, Wiley, 1969.

J. Ziman, *Principles of the Theory of Solids*, Cambridge, 1964, sec. 6.8.

PROBLEMS 1. If a room-temperature Si semiconductor is only designed to withstand a current increase of a factor of 5 before the junction is destroyed, what is the maximum increase in operating temperature over room temperature that can be tolerated?

2. Estimate the ratio of forward to backward resistance for a 1 V bias in a Si diode using the approximations developed in the text, and the definition of dynamic resistance as dV/di. How does this ratio generally depend on bias?

3. Estimate the current density for a heavily back-biased Si diode at room temperature. Use the same kind of assumption as outlined in problem 2-6.

4. Calculate the characteristic tunneling length for an electron with $v = 1 \times 10^7$ cm/s when it strikes a 1 eV high barrier.

5. Calculate the minimum time that electrons could traverse a 50 Å junction if the semiconductor is at room temperature. How does this compare with the switching time of a tunnel diode?

6. Describe the properties that give rise to the negative resistance in tunnel diodes and Gunn effect devices.

7. Explain how a local increase in the electron density can produce a corresponding change in the electric field, as observed in Gunn effect devices.

8. What properties would you expect to be required for a semiconductor to have a maximum sensitivity of its resistance to light? Why?

9. Describe the characteristics of a light source you would use to achieve a maximum photosensitivity in a semiconductor with $E_g = 2.0$ eV.

10. In designing a solid-state television camera it is necessary to produce a device sensitive to green light. Suggest how the materials in the table of semiconductors might be used to produce information about the amount of light in the yellow-green part of the visible spectrum. What materials could be used for the blue-green region?

11. Approximately how much charge would be stored on a Si diode with a depletion region of approximately 2000 Å and a back bias of 2 Volts? How much charge would be able to flow through an external shorting circuit?

12. Would it be better to construct the diodes in the vidicon from silicon or germanium? Estimate the ratio of their "holding time" for a capacitance charge.

13. Estimate the maximum possible efficiency of a $E_g = 2.0$ eV diode for converting daylight into electrical energy. Use the approximate spectrum observed for daylight. Can you suggest the design of a diode that would improve this efficiency?

14. A germanium ($\epsilon = 16 \ \epsilon_0$) p-n junction is doped with As and P. Use the Bohr model to predict what frequency light will be absorbed in the p side and in the n side. Do you think this would be a sensitive way to detect impurities in semiconductors?

15. Estimate the current that would be generated by a 100 candlepower point source of monochromatic red light ($\lambda = 0.7 \ \mu$m) placed 1 meter from the p side of a diode ($E_g = 1.0$ eV). The diode is maintained at zero external bias and the area exposed to the light is 1 cm^2. If a diode with $E_g = 2$ eV were used what current would you expect? (A 1 candlepower source produces 1.5×10^{-7} J/s $\cdot$ cm^2 at a distance of 1 m.)

16. A solid-state radiation detector is to be constructed using a p-i-n diode. What would be the best choice of E_g for the i region in order to obtain the greatest possible precision of incident radiation energy determination. Ignoring thermal effects, would you expect the device to attain the accuracy indicated by the above consideration? Why?

17. A solid-state detector is to be designed for room temperature operation and high sensitivity to γ rays. What characteristics of the semiconductor do these requirements affect, and what do they imply would be the best choice of semiconducting material for the detector?

18. Would you expect the number of conduction electrons generated in a semiconductor detector to be proportional to the deposited energy? Discuss the ways that the energy, say of an incident energetic electron, might be expended in the intrinsic region of a detector. Compare the average energy required to produce an electron-hole pair in Ge with the gap energy.

19. Why are solid-state detectors frequently cooled to improve their signal-to-noise ratio?

20. Would it be more reasonable to construct a LED out of direct gap material than indirect gap material? Why?

21. Solid-state lasers are made from direct gap semiconductor rather than indirect gap material. Can you explain why? Can you guess what advantages the use of indirect gap materials (such as silicon) might have over direct gap materials?

THE NUCLEUS

So far, we have been taking for granted the fact that at the center of any atom there is a compact nucleus of positive charge, $+ Ze$, which contains almost all the mass of the atom in the form of protons and neutrons (collectively called *nucleons*). From the mere presence of more than one positive charge confined to the atomic center we should recognize that a new force must exist. The Coulomb force can never provide attraction between two like charges, and the gravitational force, although it is always attractive, is much weaker than the Coulomb force; therefore there must be a new force—a nuclear force—keeping the protons together in the nucleus. From the size of the nucleus, determined by Rutherford scattering to be of the order of magnitude of 10^{-14} m for heavy elements, this new nuclear force must be strong—strong enough to overcome the Coulomb repulsion between the constituent protons and also provide a net attraction between them.

In the following chapters we will look at the basic properties of the nucleus, with the major aim of understanding the physics of nuclear power. To do so we will need to build up at least a gross picture of the systematics that control the structure of the nucleus. A picture of the nucleus comparable to the Bohr model for the atom would be ideal; unfortunately no such reasonable and simple picture exists for the nucleus. The details of the nuclear force, even though not yet fully understood, are much more complex than for the Coulomb force governing the atom. We will have to make do with an even cruder picture of the nucleus than the Bohr picture provides for the atom.

9.1 THE NUCLEAR POTENTIAL A gross description of the form of the nuclear force can be built most conveniently in terms of a potential picture. Much can be learned about the nuclear potential from the results of the Rutherford scattering experiment. From the fact that no deviation from Coulomb scattering was observed down to a separation of $\sim 10^{-15}$ m, we know that there is no nuclear force, and therefore no nuclear potential beyond this range. From the existence of stable nuclei we know that at some range the nuclear force must be attractive and stronger than the Coulomb force at distances less than $\sim 10^{-15}$ m; therefore, the nuclear potential in this range must be deeper than $\sim 10^6$ V. From high-energy nucleon-nucleon scattering experiments it is found that at even closer separations the nuclear force becomes strongly repulsive. These pieces of information can be used to put together a basic picture of the nuclear potential. which is shown in Fig. 9-1.

This will be used to represent the potential due to the nuclear force between any pair of nucleons. The form and magnitude of the nuclear force, furthermore, has been found to be the same for *any* pair of nucleons, that is, p-p, p-n, and n-n; it is charge-independent. In fact the nucleon charge (or absence of it) can be thought of as just another quantum number or identifying property, like intrinsic spin. The nuclear force between two nucleons is so much larger than the Coulomb force that charge has very little direct influence on the motion of nucleons in the nucleus.

9.2 NUCLEUS–ATOM SIMILARITIES AND DIFFERENCES Because the nucleus consists of particles confined to the nuclear volume, the wave nature of the particles can be expected to be important. This leads naturally once again to quantization of total energy, angular momentum and magnetic moments, and so on, as it

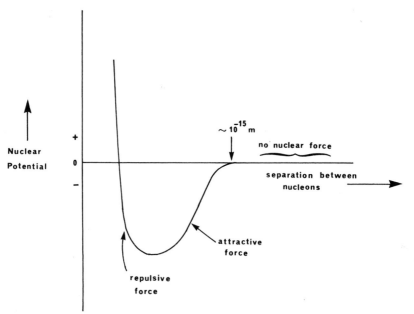

FIGURE 9-1 Basic form of nucleon-nucleon potential.

did for the atom. The proper treatment of nuclear behavior involves the solution of a Schrödinger equation, using the *nuclear* potential energy, and taking into account the Pauli exclusion principle. However here the close parallel to the atom ends. In the nucleus, unlike the atom, there is *no* strong attraction center, or "nucleus," so that the main force in the nucleus is more like the interaction between the electrons themselves in the atom. In part because of its complexity, we ignored this problem in the atom. It cannot be ignored in the nucleus. Also the nucleus consists of *two* types of particles (protons and neutrons) instead of just one in the atom (the electron). Protons have a mass approximately 1800 times that of the electron ($m_p = 1.67 \times 10^{-27}$ kg), a charge equal to the electron but opposite in sign, and an intrinsic spin the same as the electron ($s = 1/2$). Notice that this means that they have a Bohr magnetic moment,* but reduced by the electron-proton mass ratio. The neutron has a mass approximately equal to the proton, the same spin as the proton ($s = 1/2$) but no charge.

Despite the differences between the active constituents of atoms and nuclei, there are underlying similarities in the dynamics of the two systems. These similarities can help us to understand some of the systematic behavior observed in nuclei. In fact the shape of the nuclear potential is strongly reminiscent of the total energy curve for atomic molecules, and should lead us to anticipate that nuclei have at least a few molecular properties. This is indeed the case and there exists a nuclear model—the liquid drop model—based on the similarity between nuclear and molecular potentials, which pictures the nucleus as a "macromolecule." This macromolecule should be capable of rotational and vibrational excitations and have compressibility properties analogous to condensed atomic matter. Such properties, in fact, are found in many nuclei.

Appealing though it may be, the liquid drop model is unable to provide quantitative descriptions of many other properties of a wide range of nuclei, and there is a need for other "models" or starting points to quantitative understanding of the nucleus. One such starting point is to regard nucleons as being only loosely bound within the nucleus and therefore acting much like a gas of particles confined by their mutual attraction to the volume of the nucleus, but otherwise independent of one another (much like conduction electrons in a solid). Since nucleons obey the Pauli exclusion principle, the behavior of the gas cannot be as simple as that of a classical gas, and is given the name of a *Fermi gas*. Much of the picture we will build up of nuclei is based on this Fermi gas model (although not always in an easily recognized way).

A third nuclear model that has had much (initially unexpected) success is the *shell model*. This assumes that the net effect of the individual nucleon-nucleon interactions can be approximated by an overall central attraction, allowing a treatment of the nucleus that is analogous to the quantum model of electrons in a single atom. One should therefore expect to find "closed shell" nuclei—that is, nuclei that require considerably more energy to raise into excited states than others—and other nuclei

---

* This enables magnetic resonance measurements (called *N*uclear *M*agnetic *R*esonance) to be carried out for molecular hydrogen, using its *nuclear* magnetic moment.

whose excitation properties are determined mainly by a few "valence" nucleons. The shell model has had surprising success at predicting many excited state properties of a large number of nuclei. Indeed *all* models of the nuclei (these three are not the only ones) are successful under some circumstances, but *no* model is capable of reliable predictions for all nuclei. The nucleus is just too complex when looked at in any detail.

9.3 MASS-ENERGY EQUIVALENCE AND BINDING ENERGY   In the structure of atoms we found that the binding energy of the simplest system—atomic hy-

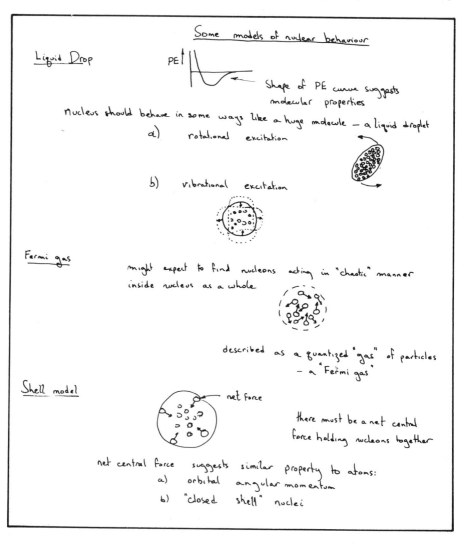

drogen—was of great use in determining energy scales. We can obtain analogous information about nuclear binding by making the simplest nuclear system we can think of. This is one neutron bound to a proton which forms a deuteron.

$$p + n \rightarrow d$$

We would expect that the mass of deuteron should simply be the sum of the proton and neutron masses. These masses have been measured, in units of MeV (recall that $E = mc^2$). The actual mass values are

$$m_n c^2 = 939.552 \text{ MeV}$$

$$m_p c^2 = 938.770 \text{ MeV}$$

$$m_p c^2 + m_n c^2 = 1878.322 \text{ MeV}$$

measured $m_d c^2 = 1876.098 \text{ MeV}$

difference   2.224 MeV

and we see clearly that the masses do not add up. The combined mass of a proton and a neutron is *greater* than the mass of a deuteron. To understand what is happening we must realize that the mass-energy equivalence pointed out by Einstein says that the *total energy* is a constant, not just the mass. When the proton and neutron come close enough together to feel the nuclear potential they become bound by a certain amount of energy. This binding energy is reflected in a decrease in the mass of the composite system. When the proton and neutron are bound together to form a deuteron, a photon is emitted (Fig. 9-2), which carries off an energy equal to the binding energy, just as in the formation of the hydrogen atom. The energy of the photon (γ ray) emitted when the deuteron is formed is 2.2245 MeV. Total energy is in fact conserved.

The detailed energy conservation statement is as follows: When the proton and neutron are far apart, and have no kinetic energy

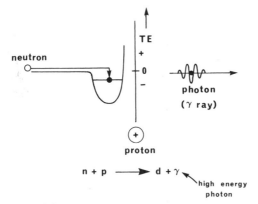

FIGURE 9-2 Formation of deuteron.

$$E = m_p c^2 + m_n c^2 + \underbrace{KE + PE}_{0}$$

$$= m_p c^2 + m_n c^2$$

When they are brought together (in such a manner that $KE \cong 0$) energy must be conserved. On forming a deuteron a photon is emitted and its energy must be included in the energy balance; the photon energy plus the deuteron mass must be the same as the total energy initially in the system.

$$E = m_p c^2 + m_n c^2 + KE + PE + E_{ph}$$

The term $KE + PE$ now is the kinetic and potential energy of the proton and neutron bound together within the deuteron. As they form a bound system this *must* be a negative number. It is the binding energy (BE) of the system. The conservation of energy then states that

$$m_p c^2 + m_n c^2 = \underbrace{m_p c^2 + m_n c^2 + \overbrace{KE + PE}^{\text{net negative}} + E_{ph}}_{m_d c^2}$$

so

$$m_p c^2 + m_n c^2 = m_d c^2 + E_{ph}$$

The binding energy, which in this case is released as a gamma ray, gives us an estimate of the energy involved in the energy levels in nuclei. It is given by

$$BE = m_p c^2 + m_n c^2 - m_d c^2$$

$$= 2.224 \ MeV$$

The very great magnitude of the nuclear force, or equivalently the nuclear potential, means that considerable amounts of mass can be converted directly to energy. This is what is exploited in nuclear power. To see *how* it can be exploited we will need to form some idea of nuclear stability and decay.

9.4 NUCLEAR STABILITY; GAMMA AND BETA DECAY   We know that the behavior of protons and neutrons in the nucleus is controlled by the detailed nature of the nuclear force, and the energy levels will be described by appropriate sets of quantum numbers. Since we have made no detailed description of the nuclear force, we cannot expect to describe in a detailed way what the energies and particular quantum numbers of the nuclear levels will be. However we can build up a picture that we know cannot be wrong if we make use of only the most basic of quantum properties to specify level properties. We know that whatever other properties nucleons have in the nucleus, they at least have spin and *must* obey the Pauli principle. Therefore we should be able to construct a basic picture of nuclear levels by assuming that each level contains at most two protons and/or two neutrons. The detailed spacing of these levels will depend on information we do not have at the moment, so, *faut de mieux* we will assume that they are spaced approximately uniformly. Whatever the actual spacing may be, we expect level energies and quantum properties to be identical for protons and neutrons, since the nuclear force is independent of particle type.

The nucleus, like the atom, when initially assembled de-excites to the lowest possible total energy state and then becomes stable against further de-excitations (Fig. 9-3). We have already seen that the de-excitation energy emitted in forming the stable deuteron is carried off in the same way as for atoms, that is, by photons. (Photons from nuclear de-excitations are normally called γ rays.) However we immediately encounter a question in the formation of stable nuclei. Do protons and neutrons separately fall into their own lowest energy states, or do they fall into the lowest energy state available to either? In other words do we treat protons and neutrons as identifiably separate, immutable particles within the nucleus or should we think of them as really identical particles, differing only by a charge "quantum number" which is interchangeable between a charge "on" state (protons) and a charge "off" state (neutrons)?

If protons and neutrons really are separate and immutable particles then there should be stable nuclei corresponding, for example, to the two mass 12 systems shown in Fig. 9-4. Both $^{12}_{6}C$ and $^{12}_{5}B$ should exist in nature as stable nuclei since their protons and neutrons are separately in the lowest energy states available to them. However if we assume that the nuclear charge is "only" a quantum number and that protons and neutrons are really interchangeable, then we can confidently predict that the configuration in Fig. 9-4a is just an "excited state" of that in Fig. 9-4b, and is therefore unstable, and will eventually decay. (The probability of de-excitation, since the nucleus is a quantum mechanical system, behaves in the same basic manner as photon emission in atoms, i.e., exponential decay.) Experimentally what is observed is that

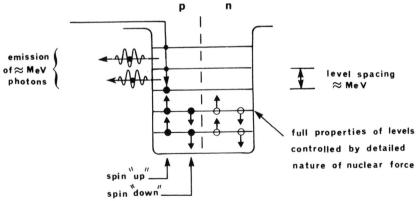

FIGURE 9-3 De-excitation of nucleus by the emission of γ rays.

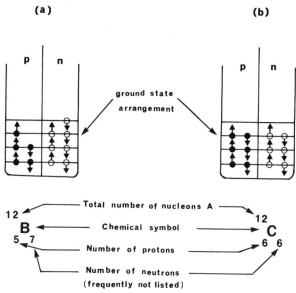

FIGURE 9-4 Lowest energy state for ^{12}B (*a*) and ^{12}C (*b*).

$^{12}_{5}$B decays to $^{12}_{6}$C, releasing 13.4 MeV. This new decay mechanism is called *beta decay*.

What is the nature of the radiation that is emitted in this decay? It must carry off a definite energy, amounting to the mass difference between (in this case) ^{12}B and ^{12}C. It must *also* carry off a charge "quantum" equal to the proton charge but opposite in sign. In fact an energetic electron, called a β ray, is emitted. However two properties of the β ray require an additional particle to be emitted. The electron has spin 1/2 which, if it were the only thing emitted, would require either ^{12}B or ^{12}C to have a half-integral ground state spin, which is not true; they both are known to have integral spin. Also, the emitted electron does not *always* carry off the same amount of energy which, if it were the only particle emitted, would indicate nonconservation of energy for the decay in general. In fact an additional particle, called a *neutrino* is emitted. This particle has spin 1/2 (allowing angular momentum to be conserved), energy (allowing total energy to be conserved), zero charge, and zero rest mass (since the maximum electron energy observed equals the mass difference of ^{12}C and ^{12}B). The neutrino, not surprisingly, is difficult to observe since it has no charge and no rest mass, and it was not until many years after the postulation of its existence that it was "observed."† Nevertheless, it nicely completes the picture of β decay.

Once having admitted that ^{12}B must be an excited state of ^{12}C, we realize that exactly the same argument must be made for ^{12}C and ^{12}N; ^{12}N must be an excited state configuration of ^{12}C, and should decay, emitting the same form of radiation as in the

† See Further Reading, Kim and Strait, sec. 14.5.

case of ^{12}B, except that a *positive* charge will be carried off. In fact ^{12}N does decay to ^{12}C, emitting a positively charged particle called a *positron,* equal in mass to the electron but with a charge equal in magnitude and opposite in sign to that of the electron. (A neutrino* is also emitted in the decay.)

We now have a basic picture of the stable configuration of a collection of A nucleons. From the assumption of equivalence of protons and neutrons there should be one stable combination of protons and neutrons for a given total number of nucleons, and that is when the number of protons, Z, equals the number of neutrons, N. Nuclei with fewer protons than neutrons decay to stability by emitting a negatively charged (β^-) particle, and nuclei with fewer neutrons than protons decay by emitting a positively charged (β^+) particle. It can be shown that the variation of total energy with Z (Fig. 9-5) is a parabola for our assumption of a constant level spacing.

This simple picture is found to become somewhat more complicated for even-A systems. These systems can be made from a combination of either even Z and even N (even-even) nuclei, or odd Z and odd N (odd-odd) nuclei. It is found that even-even nuclei systematically are bound more tightly (by approximately a constant amount)

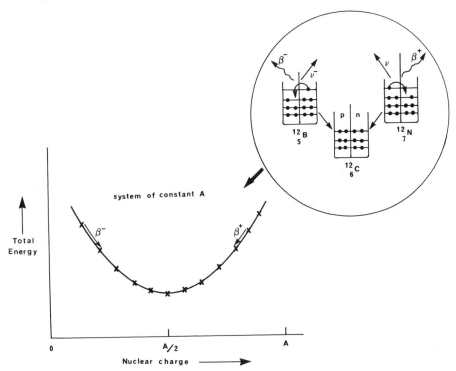

FIGURE 9-5 Total energy vs Z for a constant A.

* For reasons of interest in fundamental physics, the neutrino emitted in β^- decay is called an antineutrino, and the neutrino emitted in β^+ decay is called a neutrino.

than they should be in our simple picture. This effect can be thought of as a "residual" interaction between the least-bound pair of nucleons which gives them additional binding energy. This is roughly analogous to the Coulomb interaction between electrons in atoms that we never bothered with (for nucleons it is attractive, for electrons it is repulsive). The result of this effect is that for even-A nuclei the total energy vs Z curve really consists of two parabolas (Fig. 9-6), one for even-even nuclei and one for odd-odd nuclei where the parabolas are separated by the energy lost in breaking a nucleon pair. This makes the β stability curve somewhat more complicated, and permits more than one stable nucleus for a given even A, and the possibility of a nucleus decaying either by β^+ or by β^- emission.

The same picture shows that there should be *one* parabola of total energy vs Z for odd A nuclei. Since the minimum of the parabola occurs at $Z = A/2$ and therefore corresponds to a half-integer value of Z, there should be two stable nuclei for odd A systems; one with $Z = A/2 + \frac{1}{2}$ and one with $Z = A/2 - 1/2$. This conclusion will have to be modified in the next section.

For even A nuclei, the existence of two parabolas controls the number of stable nuclides. If A is a multiple of 2 but not 4, so that the minimum of the parabolas lies on an odd Z value, then there will be two stable nuclides for that A, one with $Z = N + 2$, and one with $Z = N - 2$, as shown in Fig. 9-7. There should never be a stable odd Z, odd N nucleus in this picture. If A is a multiple of 4, then the minimum of the parabola lies on an even Z, and there should only be *one* stable nuclide for this A. (Both these predictions for the number of stable nuclides should be correct only if the energy separation of the parabolas is less than the separation between adjacent points on the same parabola, otherwise there could be more stable nuclides than predicted.) We can easily check these predictions, and shortly will do so.

9.5 GAMMA AND BETA RADIATION SOURCE APPLICATIONS The large amounts of energy carried off in both γ and β decay of excited nuclei makes possible

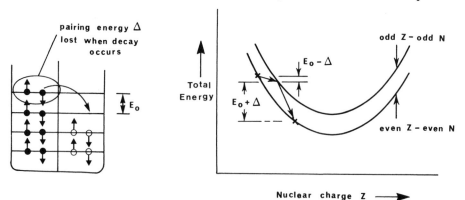

FIGURE 9-6 Effect of pairing energy on β decay of even-A nuclei.

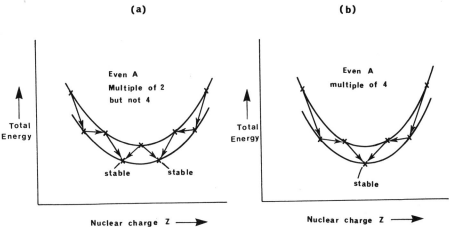

FIGURE 9-7 Minimum energy parabolas for even-A nuclei where A is a multiple of 2 but not 4 (*a*), and a multiple of 4 (*b*).

a number of practical uses of these two forms of radiation. We will discuss here one such use, thickness measurements.

Although we have not determined the detailed nature of the interaction between high-energy photons (γ rays) and matter, we know already from the basic properties of photons that the intensity of a beam of γ rays will be exponentially attenuated on passing through solid materials. The characteristic exponential attentuation coefficient depends on γ-ray energy and the nature of material it is passing through (Fig. 9-8), but characteristically is $\sim$0.5 cm^{-1} for γ rays of about 1 MeV energy. This means that γ-ray intensity attenuation measurements in general will be sensitive to material thicknesses and thickness variations up to values of the order of 1 cm or so.

The absorption of monoenergetic β rays passing through materials does *not* obey an exponential law since, unlike γ rays, the β particle interacts with every atom it passes within an atomic radius of (being a charged particle) and thus suffers a continuous energy loss. However, there are two additional factors that must also be considered. First, a source of β radiation does not emit monoenergetic β rays, but a completely continuous spectrum up to some maximum value determined by the mass difference of the nucleus before and after decay. Second, there are likely to be an appreciable number of collisions with atoms which will scatter the β rays through a relatively large angle, effectively removing them from the incident beam. These two facts have the result that when the total number of β particles of all energies emitted from the source that pass through a thickness x of some absorbing material are measured, the number left in the initial beam is found to be given to a good approximation by an exponential relation

$$N(x) = N_0\, e^{-\mu x}$$

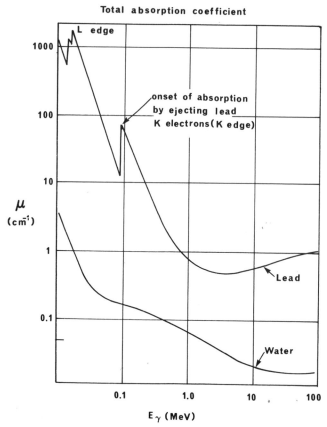

FIGURE 9-8 Total absorption coefficients for water and lead as a function of γ ray energy. Adapted from *The Atomic Nucleus* by R. D. Evans. Copyright © 1955 by McGraw-Hill Book Company. Used with permission of McGraw-Hill Book Company.

The value of μ is found to be given* approximately by the relation

$$\mu(cm^{-1}) = \frac{17\rho}{E_{max}^{1.14}}$$

where E_{max} is the maximum β-ray energy emitted in the decay and ρ is the density of the absorbing material in gm/cm³. In aluminum this would correspond to an attenuation of an incident beam of 3.5 MeV maximum energy β rays to 1/e of its initial intensity in about 0.1 cm, so β rays are generally used to measure material thicknesses and thickness changes of the order of a mm or so.

The mean life for γ emission from nuclei is typically 10^{-16}s and can vary several orders of magnitude from this. Beta-decay lifetimes are characteristically much longer, rang-

* See Further Reading, Oldenberg and Rasmussen, sec. 17.3.

ing typically from fractions of a second to several tens of years. Therefore the existence of radioactive materials in nature, or the possibility of artificially generating them, means that they can be used to provide an approximately constant source of β radiation. Since several of the available β sources decay to γ-unstable states of the final nucleus, they can also be used to provide γ-ray sources, which would otherwise not be available due to their very short γ-decay lifetime.

Beta- and γ-ray attenuation thickness monitors have been used successfully to monitor and control thin film thicknesses of various materials as illustrated in Fig. 9-9. By choosing the energy of the radiation source, the attenuation length of the radiation in the material that is being monitored can be selected so that small thickness changes give a measurable change of radiation intensity in the monitoring detector.

9.6 NUCLEAR SATURATION AND THE COULOMB FORCE

The picture of nuclear stability and β decay we have developed is based on the assumption that the Coulomb force plays no part in nuclear energetics, or at least only a very small part. Is this assumption always valid? Clearly the Coulomb repulsion energy will increase with the number of protons in the nucleus. How does the effect of the nuclear force— as shown by the binding energy—change with nucleon number? Is the nuclear binding between nucleons always much greater than the effects of Coulomb repulsion between protons?

To look at the expected variations of nuclear binding energy as a function of nucleon number we will make use of the basic shape of the nucleon-nucleon potential energy curve. As shown in Fig. 9-10, we will take as the characteristic of the nuclear force only that it has no effect beyond the range r_a, that nucleons when attracted can come together no closer than a distance corresponding to an internucleon radius of r_0, where the repulsive force takes over, and they will be bound together by an energy BE, which is called the binding energy. From the binding of the proton and neutron in the

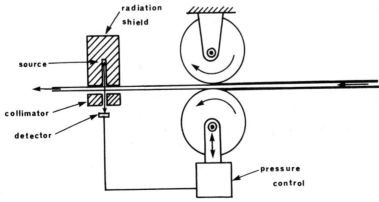

FIGURE 9-9 Schematic diagram of an on-line thickness monitor using a radioactive source.

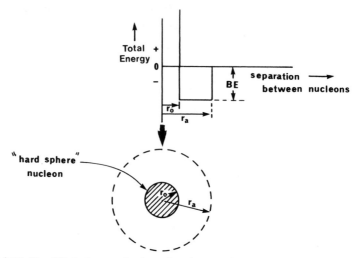

FIGURE 9-10 Simplified characterization of nucleon-nucleon potential.

deuteron, we will take the magnitude of BE to be about 2 MeV. To estimate the binding energy of a nucleon in a nucleus of A nucleons we can take a "test" nucleon of size r_0 and attractive range r_a in the middle of the nucleus and calculate its total binding energy for some assumed r_a and r_0. For ease of calculation we can take a two-dimensional nucleus, and let $r_a = 1.5\, r_0$.

As nucleons are added around the "test" nucleon, its binding energy will increase with the number of bound pairs it can make until the point is reached where it is no longer possible to add more nucleons that are nearer to it than r_a. Then no further binding energy increase will occur for the test nucleon (Fig. 9-11) when yet more nucleons are added. The binding energy is said to have become *saturated*. Clearly where this saturation occurs will depend quite strongly on the relative value of r_a and r_0; if r_a were much greater than r_0 the nucleon would "sense" many more nucleons in the nucleus and be bound more tightly. However the existence of a finite attractive *and* repulsive range for the nuclear force guarantees that *at some mass number* saturation *must* occur.

We will want to compare this behavior with what actually is observed in nuclei. However, experimentally it is not possible to isolate a nucleon that exists at the center of the nucleus (as our test particle does), but what *can* be done is to measure the total binding energy of all the nucleons in a nucleus, divide by the number of nucleons, and obtain an experimental value for the average binding energy per nucleon. Therefore it is this quantity that we should estimate and compare with experiment. The major effect of including all nucleons in our estimated curve is to round off the sharp break in the BE curve since there are now nucleons at various distances from the edge, and to somewhat reduce the rate of increase of BE vs A in the nonsaturated region. Also, the fact that real nuclei are three-dimensional, not two-dimensional, means that saturation will be reached at a different value of A than we have estimated.

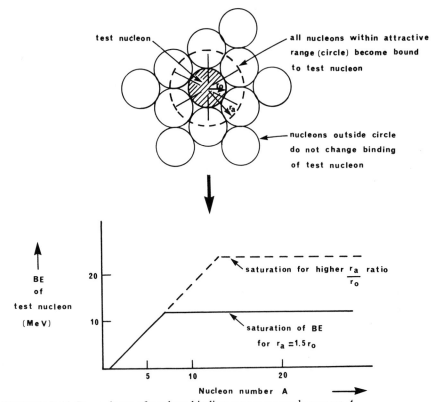

FIGURE 9-11 Dependence of nucleon binding energy on nucleus mass A.

There is one more effect that we must discuss before we actually make a comparison with experiment. The β-stability calculation indicates that approximately half the nucleons in a stable nucleus will be protons. These have a mutual Coulomb repulsion, which must reduce the binding energy and (unlike the nuclear force) since the Coulomb force has an infinite range, the Coulomb energy does not saturate but becomes continually greater. This energy can be estimated roughly for the nucleus by assuming a uniformly charged sphere of radius R. The Coulomb energy of such a sphere is given by

$$E_{\text{Coul}} = \frac{3}{5} \frac{kZ^2e^2}{R}$$

where R is the radius of the nucleus. Now the nucleus we are considering consists of A spheres of radius approximately r_0. Hence its radius is simply related to the nucleon radius and nucleon number

$$\frac{4}{3} \pi R^3 \cong A \times \frac{4}{3} \pi r_0^3$$

where r_0 is the nucleon "size" (found to be about 1.3×10^{-15} m), so that

$$R = A^{1/3} \, r_0$$

(This is found to be reasonably obeyed experimentally.) Since the nucleus is approximately half protons we obtain (for large A)

$$
\begin{aligned}
E_{\text{Coul}} &= \frac{3}{5} \frac{k \left(\dfrac{A}{2}\right)^2 e^2}{A^{1/3} r_0} \\
&\cong \frac{3}{20} \times \frac{9 \times 10^9 \, A^2 \times (1.6 \times 10^{-19})^2}{(\sim 5) \times 1.3 \times 10^{-15}} \\
&= 0.02 \text{ MeV } A^2
\end{aligned}
$$

The repulsive Coulomb energy per nucleon, which we require to correct our average binding energy per nucleon curve, is then simply

$$\frac{E_{\text{Coul}}}{A} \cong 0.02 \, A$$

The expected total average binding energy per nucleon curve should be corrected for this repulsive term, and will therefore look qualitatively like the one shown in Fig. 9-12(a).

The experimental binding energy curve clearly confirms our basic picture. With a few (notable) exceptions the average binding energy per nucleon increases with increasing nucleon number until saturation is reached. This occurs around mass 50. After saturation has occurred the binding energy per nucleon decreases more or less linearly with nucleon number, in a fashion consistent with our description of Coulomb repulsion. The slope of the curve in this region is somewhat less than our estimate, -0.008 MeV/A compared with our value of -0.02 MeV/A. This discrepancy is not serious in view of the simplicity of our calculation (for instance, we have overestimated the number of protons in using our initial prediction).

The deviations from the smooth curve are due to the important nuclear force details that we have ignored. In fact the extra-strong binding of ^{4}He, ^{16}O, ^{20}Ne, and so on are indications of a shell-like structure. We will make use of the especially strong binding of ^{4}He (also called the α particle) shortly.

The fact that the Coulomb repulsion has an appreciable effect on nuclear binding for high-A nuclei will make a change in our previous decision about the neutron-proton composition stable nuclei. Our total energy vs Z curve must also be modified to show the decrease of binding energy due to Coulomb repulsion. The effect is shown in Fig. 9-13. This correction becomes more important as A (and therefore, Z) increases.

In general the Coulomb repulsion reduces the binding energy of the nucleus (decreases the negative total energy), as shown in the average binding energy per nucleon curve, and also shifts the proton/neutron ratio in heavy nuclei towards a neutron excess, by

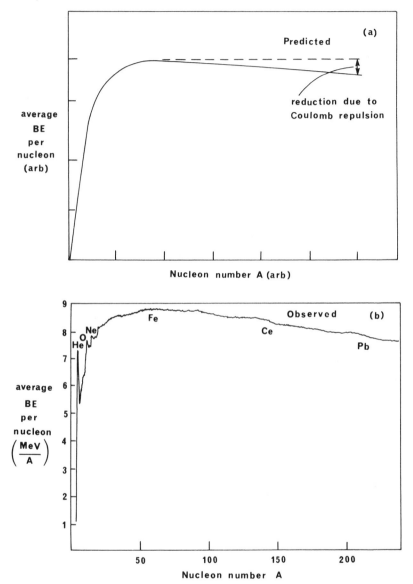

FIGURE 9-12 Predicted (*a*) and observed (*b*) mass-dependence of average binding energy per nucleon.

an amount that will *increase* with increasing A. A significant shift can be expected only after the saturation of the nuclear force has occurred, that is, for $A \geqslant 50$.

The observed stable nuclei as a function of Z and N are plotted in Fig. 9-14. For low mass the stable nuclides cluster closely about the $N = Z$ line (as anticipated by the initial picture); for nuclei much beyond the saturation point of the binding energy

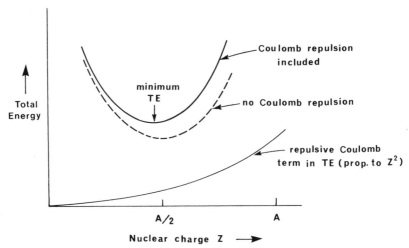

FIGURE 9-13 Effect of Coulomb repulsion on β stability curve.

curve, the stable nuclides leave the $N = Z$ line and all have a neutron excess, consistent with the increasing importance of Coulomb repulsion. In addition the effect on odd A nuclei is such that there will be only *one* stable nuclide per value of odd A in general since the Coulomb distortion shifts the position of the parabola. This is in fact what is observed. With the exception of ^2_1H, ^6_3Li, $^{10}_5B$, and $^{14}_7N$ there are no stable odd-odd nuclei, as expected. The exceptions are all very light nuclei, and their properties have to be obtained using a more detailed treatment of the nuclear force* than the extremely simple one we have employed. However, the general picture is nicely confirmed by the known stable nuclei.

9.7 THE OVERALL GROUND STATE OF NUCLEAR MATTER   To understand the instability of nuclei against β decay (the change of Z) we used a plot of total energy vs Z. The average binding energy vs A curve, when inverted (Fig. 9-15), bears a striking resemblance to that plot and strongly suggests that the "ground state" of overall nuclear matter—where all nucleons are bound by the greatest amount possible—should be nulcei in the mass region of iron. (Experiment shows that ^{576}Fe is the most tightly bound nucleus.)

We should be able to increase the binding energy of nucleons, and release the excess energy in some form of radiation by combining light nuclei and by splitting heavy nuclei. Such is indeed the case; the former process, called fusion, releases energy by increasing the binding energy due to the nuclear force. The latter process, called fission, releases energy by decreasing the Coulomb repulsion within the nucleus. The fission process at present is the only one from which we are able to extract usable energy. Therefore we will first look at the physical process that can reduce the number

* See Further Reading, deBenedetti, sec. 2.13.

Stable Nuclides

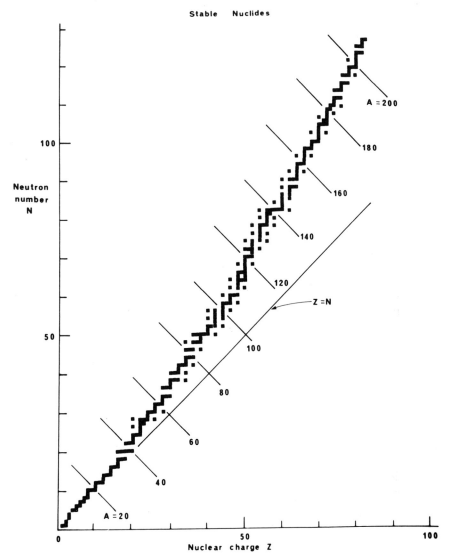

FIGURE 9-14 Plot of stable nuclides as a function of N and Z.

of nucleons in the nucleus and release energy. In the next chapter we will look at some of the properties and difficulties of systems necessary to control the process of energy extraction.

9.8 NUCLEON DECAYS OF NUCLEI In order to decrease the mass number of a nucleus, one or more nucleons have to be emitted. At first glance it might appear

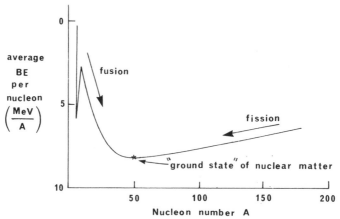

FIGURE 9-15 Inverted average binding energy per nucleon curve, showing de-excitation via fission and fusion.

that this would be energetically possible for any nucleus past the peak of the binding energy curve. However we must remember that the curve as drawn represents an *average* energy, and what must be gained is *total* energy. For nuclei past the peak in the binding energy curve there will be a net gain in binding energy of approximately the binding energy curve slope times the number of nucleons emitted, but this gain will be reduced by the binding energy loss of the emitted particle(s). If there actually is a net binding energy gain to be had by emitting a nucleon, then the process can (and will) occur, even though it may proceed at a very slow rate (we will discuss decay rates later).

At what nucleon number should the nucleus be unstable to the emission of a single nucleon?

$$A \rightarrow (A - 1) + 1$$

$$BE \text{ gain} = (A - 1)\Delta E - 1 \times \frac{BE}{A}(A)$$

The quantity ΔE is the binding energy gain per nucleon, and numerically is equal to the slope at nucleon number A of the binding energy curve. The binding energy curve is denoted BE/A and is a function of the nucleon number A. In the region we are concerned with, the experimental binding energy curve can be fitted by the straight line

$$\frac{BE}{A} = [9.5 - \overset{\overset{\displaystyle \Delta E}{\frown}}{0.008A}]$$

We can insert this into the relation for the binding energy gain to see at what value of A this becomes a positive number (the nucleus just becomes unstable to single-nucleon decay). We will denote this threshold value of A by A_{th}. The relation is

$$0 = (A_{th} - 1) \times 0.008 - 1 \times (9.5 - 0.008A_{th})$$

$$9.5 \cong 0.016A_{th}$$

$$A_{th} \cong 594$$

According to this calculation, the nucleus should not become unstable to single-nucleon emission until a nucleon number of ~ 600 is reached. However there are no known stable nuclides beyond a mass number of about 240, so that this process cannot be contributing to the decay of any existing nuclei nor can it be the cause of the nonexistence of nuclei beyond mass 240.

The reasons that nuclei are not unstable to single-nucleon emission until such large mass numbers are twofold. First and most important is the fact that several MeV of binding energy is lost in the emission process because the emitted single nucleon is not bound to anything. Second, the change in mass number by only one increases the average binding energy of the remaining nucleons by only a very small amount. In order for nucleon emission instability to occur at lower mass numbers we must consider the emission of clusters of nucleons.

The cluster that immediately looks most favorable for emission is ^{4}He, the α particle. It has a very high binding energy per nucleon, alleviating the first problem, and will quadruple the binding energy gain per nucleon of the residual nucleus, helping the second problem. Since there are relatively few nucleons (4) that must be clustered together within the nucleus the probability of an alpha particle appearing at the surface of the nucleus can be relatively high, although much lower than for single nucleons. The energy condition for decay by α emission is

$$A \rightarrow (A - 4) + 4$$

$$BE \text{ gain} = (A - 4)(4 \times 0.008) - 4\left[\frac{BE}{A}(A) - \frac{BE}{A}(4)\right]$$

The value of A where the onset of a net gain of energy occurs, A_{th}^α, is given by

$$0 = (A_{th}^\alpha - 4)(4 \times 0.008) - 4[(9.5 - 0.008\, A) - 7.1]$$

$$= (A_{th}^\alpha - 5)(4 \times .008) - 9.6$$

$$A_{th}^\alpha = 152$$

In fact the first naturally occurring nucleus unstable against α decay is ^{142}Ce, which decays to ^{138}Ba, emitting a 1.5 MeV α particle. However such decays do not occur very frequently. Any individual ^{142}Ce nucleus has a 50 percent probability of having α-decayed only after 5×10^{15} years (i.e., the half life of ^{142}Ce is $T_{1/2} = 5 \times 10^{15}$ years)! There are basically two reasons for this low probability. First the combined motion of all the nucleons in the nucleus must be such that they act like a ^{138}Ba nucleus and an α particle moving away from each other. The chances of this occurring at any given time are small. Second, and most importantly, when this does occur the

α particle must penetrate, or tunnel through, a large potential barrier as shown in Fig. 9-16. Because both ^{138}Ba and the α particle have a charge there is a Coulomb potential between the two which, together with the attractive nuclear force, forms a potential barrier, commonly called a Coulomb barrier. The magnitude of this barrier is quite sizable, so that the α particle has to tunnel a considerable distance.

The magnitude of the Coulomb energy barrier at the outside "edge" of the ^{138}Ba nucleus is approximately

$$E_{\text{Coul}} \cong \frac{k\,Z_1 Z_2\,e^2}{R_{Ba}}$$

$$\cong \frac{9 \times 10^9 \times 56 \times 2 \times (1.6 \times 10^{-19})^2}{(138)^{1/3} \times 1.3 \times 10^{-15}}$$

$$\cong 25\ \text{MeV}$$

In fact the α particle must tunnel out to about 100×10^{-15} m before it has penetrated the Coulomb barrier. Remembering that the penetration probability for a particle depends exponentially on barrier width and particle energy, it is not surprising that the decay rate is very low for ^{142}Ce. A detailed calculation of the penetration probability and hence α-decay rate using a Coulomb potential has been worked out as a function of nuclear charge and decay energy.* The observed systematic energy dependence is shown in Fig. 9-17.

What should the systematic variation of α decay probability with increasing mass number be? A major controlling factor (ignoring considerations of α formation probability) is the relative Coulomb barrier height and excess energy available. The relative

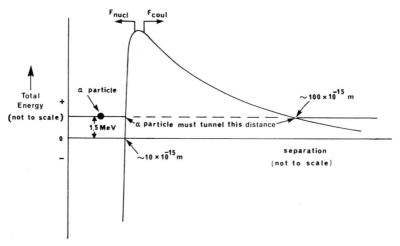

FIGURE 9-16 Barrier penetration in ^{142}Ce α decay.

* See Further Reading, Leighton, sec. 15-6.

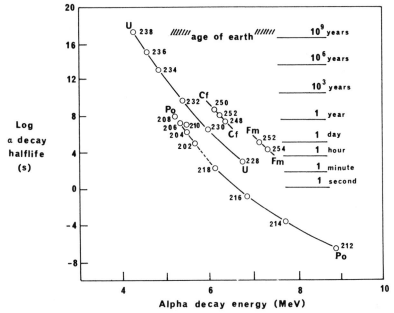

FIGURE 9-17 Plot of alpha-decay half life against decay energy for several alpha emitters. Besides the expected strong dependence on alpha-decay energy, the half lives depend strongly on nuclear size and internal structure properties.

importance of these two sets of factors can be seen from the information in Fig. 9-17, where a roughly similar set of nuclei show a variation of $\sim 10^6$ in half life due to internal nuclear properties, but a variation $\sim 10^{25}$ due to alpha energy variation. We can see from our binding energy gain calculations that the excess energy available increases linearly with A; the Coulomb barrier height also increases with A, but not quite so rapidly,

$$E_{\text{Coul}} = \frac{kZ_\alpha Z_{(A-2)}}{R} \cong \frac{2kA/2}{r_0 A^{1/3}} = \text{const} \times A^{2/3}$$

so that instability against α decay becomes greater as the nuclear mass increases. In fact this process is the cause of the limit of the main stable nuclide line at $A \cong 210$. Beyond that value *all* nuclei are α-unstable, with measurably short half lives*. We might ask ourselves after all this why we saw α decay in ^{142}Ce in the first place. It should occur, even within the errors of our crude calculation, with a negligible α particle kinetic energy, and therefore have an immeasurably long half life. The answer to this seems to be that there is a little hump in the *BE/A* curve (due to nuclear structure effects) which gives the ^{142}Ce the extra decay energy necessary for the process to be observed.

*The half life, $T_{1/2}$, is the time required to reduce the number of excited nuclei by a factor of two. Therefore $T_{1/2} = \tau \ln 2$, or 0.69τ

9.9 LARGE CLUSTER DECAY AND FISSION From the arguments used in discussing α-decay instabilities, that is, the high binding energy of the α particle and the large per-particle energy gain of the remaining nucleus, we might expect the decay of nuclei by the emission of larger fragments, such as oxygen and neon, to be even more favorable. Certainly such decays will be allowed energetically at lower masses, but there are two additional factors that militate heavily against such processes. The Coulomb barrier increases with the charge of the fragment roughly as

$$E_{\text{Coul}} = \frac{kZ_1Z_2}{R}$$

$$\cong \text{const} \times Z_1,$$

for a fixed value of A of the decaying nucleus, decreasing the penetration probability severely. Here we have assumed that Z_2 is roughly constant since Z_2 is $>> Z_1$. More importantly, however, the probability of forming large clusters inside the nucleus decreases drastically with the number of nucleons in the cluster. Neither of these facts *strictly* forbid the breakup of a heavy nucleus into a lighter one plus a fragment with $\approx$ 10 to 20 nucleons in it; they just indicate that the process is highly unlikely, and one would have to wait for a rather long time before it is likely to happen. Here a long time is something much greater than 10^{15} years!

If these factors are re-examined for very heavy fragments, however (say on the order of half the total mass of the decaying nucleus), the situation changes. The increase of the Coulomb barrier slows down as a function of Z_1 for a given A since Z_2 can no longer be assumed to be even roughly constant. This, coupled with the increased gain in binding energy with the number of nucleons emitted in the cluster, or fragment, enhances the probability of penetrating the barrier. Even more importantly, the correlation of the behavior of the large number of nucleons necessary to produce two approximately equal fragments moving away from each other becomes much more probable.

This happens because the motions of individual nucleons required for the nucleus to act as two mutually separating clusters are virtually the same as those that occur in vibrations of the nuclear surface (Fig. 9-18), which for heavy nuclei is a highly likely mode of behavior. We should not be surprised to find that this type of behavior exists in nuclei, when we remember that the shape of the nucleon-nucleon potential is very similar to the atom-atom potential which led to molecules, which exhibit rotational and vibrational behavior.

The combination of the two enhancing factors (increased barrier penetration and fragment formation probability) is sufficient to allow the spontaneous breakup, or fission, of a nucleus into two roughly equal fragments at a measurable rate for nuclei in the mass 230 $\rightarrow$ 240 region. The nucleus ^{235}U is the first naturally occurring nucleus that undergoes measurable spontaneous fission. The binding energy gain for this process can be estimated quickly.

small – amplitude vibrations

(light nuclei)

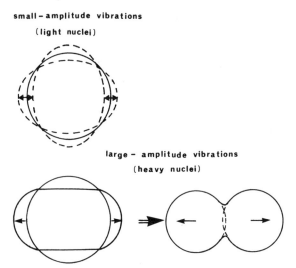

large – amplitude vibrations

(heavy nuclei)

FIGURE 9-18 Surface vibration in heavy nuclei, leading to the possibility of fission.

$$BE \text{ gain } (^{235}U \text{ fission}) \cong 2 \times \frac{A}{2} \frac{BE}{A} (A/2) - A\frac{BE}{A} (A)$$

$$\cong 235(8.4) - 235(7.6)$$

$$\cong 190 \text{ MeV}$$

This large amount of energy nevertheless is about 5.5 MeV less than the Coulomb barrier height for the two fissioning fragments, so that they must tunnel through the barrier once they have formed. Of course ^{235}U is also α-unstable, with a half life of 7×10^8 years; approximately 3 in every 10^{10} decays of ^{235}U is due to the fission process, an exceedingly small number, but nevertheless measurable.

The actual progression of fissioning nuclear matter towards its final stable state is complicated by the fact that after fission the fragments (whose nucleon number ranges anywhere from $\sim A = 70$ to $A = 160$) have a neutron-proton ratio approximately the same as that of ^{235}U. The occasional neutron may be emitted directly in the fission breakup, but this is certainly not enough to alter sensibly the decay product neutron excess, which will be very large, as shown in Fig. 9-19. This means that the decay products must significantly reduce their N to Z ratio before they can become stable; they are therefore highly unstable to β^- decay. Many of the fragments are excited so highly that they are actually unstable against neutron decay, so that a chain of direct neutron emission and subsequent β decays occurs as the fragments de-excite to their stable Z and N configuration.

The total excess energy released by fission consequently is divided between kinetic energy of the fission fragments and neutrons, and energy of the subsequent decays.

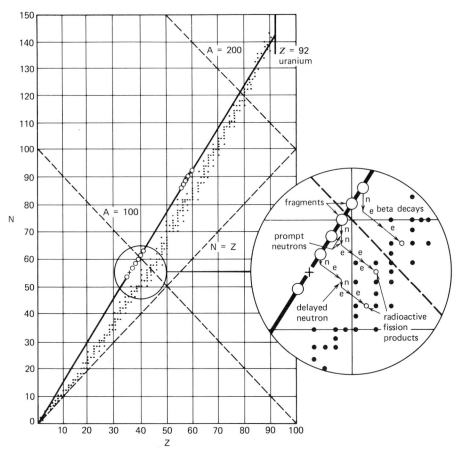

FIGURE 9-19 Plot of stable nuclides as a function of Z and N, showing composition and decay of fission fragments. From Inglis, *Nuclear Energy: Its Physics and its Social Challenge,* Fig. 4-13. Copyright © 1973, Addison-Wesley, Reading, Massachussetts. Reprinted with permission.

The approximate division of the energy for ^{235}U fission is indicated in the following table.

TABLE 9-1 Distribution of Fission-Decay Energy

Mode	Energy (MeV)
Fragment kinetic energy	167
Neutron kinetic energy	5
γ rays (fission)	5
β decay energy	5
β-induced γ rays	5
Neutrinos	11

Of the total energy release almost all of it is available immediately on fissioning; a relatively small amount ($\lesssim 20$ MeV) is produced at a later time in β-decay processes, and only the energy carried off by neutrinos emitted in the β decay is completely lost.

FURTHER READING S. deBennedetti, *Nuclear Interactions,* Wiley, 1964.

R. D. Evans, *The Atomic Nucleus,* McGraw-Hill, 1955.

S. K. Kim and E. N. Strait, *Modern Physics for Scientists and Engineers,* Macmillan, 1978.

R. B. Leighton, *Principles of Modern Physics,* McGraw-Hill, 1959.

J. Norwood, Jr., *Twentieth Century Physics,* Prentice-Hall, 1976, Ch. 11.

O. Oldenberg and N. C. Rasmussen, *Modern Physics for Engineers,* McGraw-Hill, 1966.

J. M. Reid, *The Atomic Nucleus,* Penguin, 1972.

PROBLEMS 1. Since they encounter no Coulomb barrier, neutrons are a very useful projectile, capable of exciting characteristic γ radiation from virtually all nuclei. Consequently they can be used to determine the amount of various elements present in some object. This process is called neutron activation analysis. A source of neutrons can be obtained by using a α source which produces the following reaction: $^{9}_{4}\text{Be} + ^{4}_{2}\text{He} \rightarrow ^{12}_{6}\text{C} + ^{1}_{0}\text{n}$. How much energy is released in this reaction?

2. Calculate the Coulomb repulsive energy between two protons at a separation of 10^{-15} m (1 fm).

3. Magnetic resonance measurements have been described using atomic hydrogen; however, most hydrogen exists in nature in molecular form where the electron spins are coupled together to give a total spin of zero, and hence no magnetic resonance effect from electron spin. Nevertheless the individual nuclei (the protons) have an intrinsic spin of s = 1/2, which can be made to behave in an analogous manner to the electron in atomic hydrogen. What would the resonance frequency for this effect be in a 1 Tesla magnetic field? Compare the frequency to that for the same effect with an atomic hydrogen electron.

4. The nucleus ^{74}As $\beta-$ decays to both ^{74}Se and ^{74}Ge (one by β^{+} decay and one by β^{-} decay). Calculate the energies available (in MeV) for these two decays. How are these energies related to the kinetic energy of the β particles emitted?

5. Using the mass tables in Appendix 3, calculate the total energy as a function of Z for the $A = 142$ nuclei. According to your calculations, what isotope(s) should be stable?

6. Show that an extra $2m_e c^2$ is required when calculating the energy release in β^{+} decay from the relevant atomic masses compared to the analogous calculation for β^{-} decays.

7. To what form of nucleon-nucleon force does the assumption of a uniform level spacing, which was used in describing the β-stability of nuclei, correspond?

8. How would you expect the steepness of the β-stability curve estimated in sec. 9-4 to change if the nuclear force could be reasonably characterized by a square well, as illustrated in Fig. 9-10? (See also problem 3-10.)

9. How would you expect the β-stability curve to vary with Z if the nuclear levels involved were characteristic of shell-model levels for large ℓ?

10. From the mass tables in Appendix 3 calculate the β-stability curves for mass 37, 67, and 101. Compare these curves with the estimates of sec. 9-4 and problems 7 and 9.

11. What properties of a neutrino and a photon are (a) identical, and (b) different?

12. In calculating the binding energy of various nuclei, the mass of neutral hydrogen is used, rather than the mass of the proton. Why should it make a difference, and why must it be done this way?

13. Use the following two-dimensional picture of the nucleon-nucleon total energy curve: (a) nucleons do not interact at separations $> r_a$; (b) for distances $\leq r_a$ nucleons attract each other to a separation r_0 ($r_a = 1.2\, r_0$) where they are bound by 2 MeV. Plot the binding energy of a single nucleon as it is surrounded by more and more nucleons. What would happen if r_a were much greater than r_0?

14. At mass 30 the average binding energy per nucleon is approximately 8.25 MeV/A. From this and the known masses of protons and neutrons estimate the mass of ^{30}Si. Compare your value with the known mass.

15. Using the mass tables in Appendix 3, calculate the average $\dfrac{BE}{A}$ for ^{56}Fe, ^{235}U, and ^{12}C. Calculate the binding energy of the least-bound nucleon for these same nuclei.

16. Estimate how near to "saturation" the nuclear force is at $A = 200$ for a three-dimensional model of the nucleus with $r_a = 2.1\, r_0$ and $r_0 = 1.5\, r_0$.

17. Give the basic reason that nuclei can be unstable to α and β decay in terms of the forces within the nucleus.

18. The nucleus $^{8}_{3}$Li is unstable. Bearing in mind the shape of the average binding energy curve name the energetically possible modes of decay to final stability. Calculate the energy released by each of all such decays. If $^{8}_{3}$Li is produced by capture of a slow ($KE = 0$) neutron and it γ-decays by a single γ emission to its ground state (assume that it does decay to the ground state before particle-decaying), what is the γ-ray energy?

19. Estimate the systematic dependence of alpha-decay energy release on the mass A of the decaying nucleus.

20. Are there any values of A less than 100 for which there are no stable nuclides? Can you think why this is so?

21. The average binding energy per nucleon for ^{12}C is 7.709 MeV/amu. Estimate the mass number at which decay by the emission of ^{12}C should become energetically possible, using the approximation to the shape of the binding energy curve given in the text. Why wouldn't you expect to observe spontaneous decay via the emission of ^{12}C?

22. Using Fig. 9-12, estimate what the lowest value of A is where fission is energetically possible.

23. If two deuterons (^{2}H) collide with enough energy to come within range of each other's nuclear potential, an alpha particle (^{4}He) can be formed. How much energy is released in this process (in MeV)? Use Appendix 3 for masses. If the "radius" of the deuteron is 1×10^{-15} m, what initial kinetic energy of the deuteron would be required to start the reaction? How much energy would be available from the ^{1}n + ^{3}He → ^{4}He reaction? How much energy is needed to initiate this reaction?

24. Calculate the fraction of the initial mass converted to energy when ^{235}U fissions into two equal-mass stable nuclei. What is the corresponding fraction when two protons and two neutrons are put together to make ^{4}He?

TEN

PRACTICAL NUCLEAR ENERGY

10.1 SPONTANEOUS AND INDUCED FISSION The 200 MeV of energy released by fission of a ^{235}U nucleus compared to 1-2eV released in a typical chemical reaction makes it clear that fission is capable of energy releases about 10^8 times greater than conventional energy sources. How can we make use of this tremendous energy source? The simplest, and therefore the first question we should ask ourselves is can we make use of spontaneous fission as a practical energy source? Let us take a manageable amount of ^{235}U and see what kind of power it produces.

For any of the types of decay or de-excitation we have dealt with—γ rays, α decay, β decay, or fission—the probability of decay per unit time per nucleus is a *constant* (whose value is determined by the quantum mechanical relations of the initial and final states of the system and the nature of the decay radiation), characterized by a mean life τ. The number of decays per unit time for a collection of nuclei is simply related to the number of nuclei present at that particular time

$$\frac{dN}{dt} = \frac{N(t)}{\tau}$$

(This relation leads to the characteristic exponential decrease of nuclei; $N(t) = N_0 e^{-t/\tau}$.) Since the half life* of ^{235}U is 7×10^8 years ($\tau = 3 \times 10^{14}$ s) we can make the approximation

$$\frac{dN}{dt} = \frac{N_0\, e^{-t/\tau}}{\tau} \cong \frac{N_0}{\tau}$$

* The half life, $T_{1/2}$, is the time required to reduce N by a factor of two. Therefore $T_{1/2} = \tau\, ln\, 2$, or $0.69\, \tau$.

for any practical period of time. The power output of a collection of N_0 nuclei will be given by

$$P = \frac{dE}{dt} = \frac{dN}{dt}(E_{\text{fis}}f_{\text{fis}} + E_\alpha f_\alpha)$$

where E_{fis} is the energy release per fission and f_{fis} is the probability that a decay will proceed via fission (rather than α decay). We will assume that $E_{\text{fis}} = 200$ MeV ($\sim 4 \times 10^{-11}$ J) and (incorrectly) for the moment that $f_{\text{fis}} = 1$. The *maximum* power we could obtain from $\sim 10^{23}$ atoms of uranium, which is a typical macroscopic amount, would then be

$$P = \frac{N_0}{\tau} E_{\text{fis}}$$

$$= \frac{10^{23}}{3 \times 10^{14}} 4 \times 10^{-11} = 10^{-2} \text{ W}$$

This is a very small value, particularly when compared with a standard car battery which can produce on the order of 100 W. It is even more discouraging when we remember that $f_{\text{fis}} \cong 10^{-10}$, and not 1! In fact the power release will be due mainly to α decays, which release ~ 5 MeV, not 200 MeV, and therefore there will be a factor of 40 less power than our estimate above for ^{235}U. Why is the power release so small, especially when compared with the gain we expected from the ratio of fission to chemical energy release? The answer, of course, is because the value of τ is so large that it takes a very long time to release any appreciable amount of energy.*

Obviously if we are to produce power in practicable amounts we cannot wait for spontaneous fission to release the fission energy. In order to circumvent the characteristic decay time it is necessary to excite the nucleus into a strong vibrational mode of motion and to do so at a total energy that is above the Coulomb barrier; that is, add about 5.5 MeV of excitation. There are several ways in which this energy can be added: photon absorption, bombardment by charged particles, or bombardment by neutrons. Of these, the collision probability for photons is very small, and charged particle collisions have a rather large Coulomb barrier to overcome. However the neutron has no barrier to overcome, so it can be absorbed by a ^{235}U nucleus at any incident energy. Furthermore simply absorbing a neutron into the nucleus increases the total energy by the neutron binding energy, which is greater than the required excitation energy so that the absorption of a neutron with essentially no kinetic energy is sufficient to induce the nucleus to fission immediately. Perhaps the most critically important factor about this particular mode of excitation is that the fission process itself provides free neutrons which themselves can be used to trigger another fission.

---

* It is possible however to obtain usable amounts of power from relatively short-lived β-unstable isotopes, since, although the energy release is less, the lifetime is so much shorter that practical power can be obtained in this way. See Further Reading, Foster and Wright, Ch. 7.

Neutron-induced fission can be made to produce a self-sustained reaction, as we will see in detail later.

10.2 CHOICE OF FUEL From the fact that the very low power level inherent in spontaneous fission can be overcome by *inducing* the fission process through neutron bombardment, and that the emission of neutrons occurs during fission, it is clear that a self-sustaining reaction that is capable of producing an essentially unlimited power output can be set up, in principle.

Up to this point we have only mentioned ^{235}U as a possible fuel. In fact there are several possible materials available (Table 10-1) although only two of them, ^{235}U and ^{238}U, occur in nature; the rest must be generated by neutron-irradiation of natural materials. Although this is in itself no barrier to using these other materials, there is a considerable amount of technology involved in obtaining and isolating the material and it would be premature to look into the details of those processes at this point. We will therefore only consider the use of "stable" fuel in investigating the physics and physical conditions involved in a fission power source, that is, a reactor.

What is the difference between ^{235}U and ^{238}U, and how might it affect our choice of reactor fuel? As ^{235}U is an odd-A nucleus, adding a neutron to it makes a neutron pair; this is not the case for ^{238}U. This means (as we found out in β-decay energies) that more excitation energy is gained by adding a neutron to ^{235}U than to ^{238}U, as is shown by their neutron capture Q values (energy excess from the reaction). In fact in ^{235}U the energy gained by the capture of essentially zero kinetic energy neutrons is greater than the fission barrier, so that fission can occur without any additional neutron energy. For ^{238}U, on the other hand, the relation between capture Q and fission barrier height is such that on the order of 1 MeV neutron kinetic energy is required to induce prompt fission.

This difference of the two uranium isotopes has a major effect on the probability of neutrons being captured and causing fission in ^{235}U compared to ^{238}U. As can be seen in Fig. 10-1, at very low neutron energies (0.1 eV or less, called the thermal region)

TABLE 10-1 Materials for Nuclear Fuel

"Stable"	Artificial	Neutron Capture Q (MeV)	Fission Barrier (MeV)	Thermal Fission?
	^{233}U	6.6	4.6	√
^{235}U		6.4	5.3	√
^{238}U		4.9	5.5	no
	^{232}Th	5.1	6.5	no
	^{231}Pa	5.4	5.0	√
	^{237}Np	5.0	4.2	√
	^{239}Pu	6.4	4.0	√

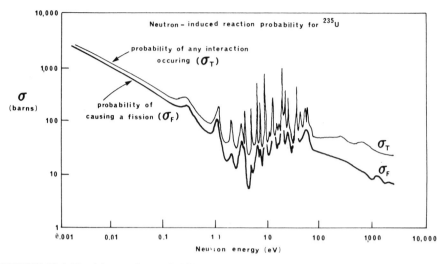

FIGURE 10-1 Total interaction probability, σ_T, and induced fission probability, σ_F, for neutrons bombarding ^{235}U, listed as cross sections. The fission cross section plot is interrupted at points where σ_F and σ_T overlap, in order to avoid confusion. Adapted From Garber and Kinsey, *Neutron Cross Sections,* vol. II, Curves, BNL-325.

the probability of fission is very large, and increases with decreasing neutron velocity. (They are going slower, so are near the uranium nucleus longer and therefore are more likely to be captured.) At these energies the probability that fission will occur, listed in Fig. 10-1 as σ_F, is very nearly equal to the total probability that any interaction at all will take place (σ_T). As the neutron energy approaches energies greater than approximately 100 eV the relative probability of fission occurring, that is, σ_F/σ_T, becomes much smaller, mainly because at these energies many other interaction processes become possible.

As can be seen in Fig. 10-2, the probability of fission at thermal energies in ^{235}U is orders of magnitude greater than that possible with ^{238}U, mainly because of the high neutron energy required to induce fission in ^{238}U, as we discussed. This factor alone strongly suggests that we will have a greater likelihood of success in designing a power system if we use ^{235}U as a fuel. Consequently we will assume that we are using ^{235}U as the basic fuel for our reactor.

10.3 NEUTRON MODERATION   The dependence of fission cross section on energy shows that the best neutron energy for operating a reactor is the lowest possible,

---

* The probability of interaction is presented as a cross section. The detailed relation between interaction probability and cross section is discussed in appendix 1. The cross section for nuclear interactions is given in units called barns (1 barn = 10^{-24} cm^2); typical nuclear interaction cross sections are a few barns.

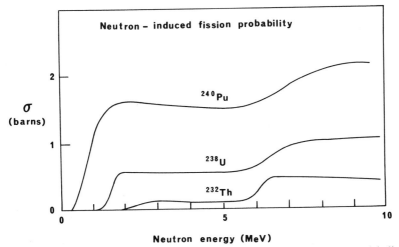

Neutron energy (MeV)

FIGURE 10-2 Neutron-induced fission probability for various even-*A* target nuclei, listed as cross sections. Adapted From Garber and Kinsey, *Neutron Cross Sections,* vol. II, Curves, BNL-325.

on the order of 0.02 eV or so. Since they come to approximate thermal equilibrium with the surrounding material it is not possible to reduce the neutrons' average kinetic energy further in any practical manner. (To reduce average neutron energies significantly below $\approx$ 0.02 eV would require refrigeration!) However, as can be seen in Fig. 10-3, the energy of neutrons emitted by the fission of ^{235}U is considerably greater than this, so we see that we must slow the neutrons down to thermal energies to make

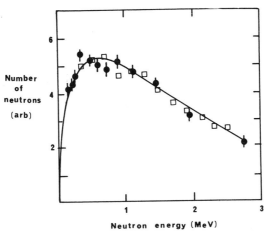

Neutron energy (MeV)

FIGURE 10-3 Energy distribution of fission neutrons from ^{235}U. The circles and squares are two different sets of experimental measurements, and the solid line is a fit to the data. From *The Physical Theory of Neutron Chain Reactions* by A. M. Weinberg and E. P. Wigner, The University of Chicago Press, Copyright © 1958 by The University of Chicago.

them most effective. How can the neutrons be slowed down to these energies? Clearly the Coulomb interaction cannot be used, since the neutron has no charge. Neutrons can only be slowed down by interacting with nuclei, that is, by "colliding" with them.

In order to most effectively lose kinetic energy by collisions, the atoms in the material used to slow them down should have as low a mass as possible. Ideally this material would be hydrogen, since a neutron can lose up to 100 percent of its kinetic energy to a hydrogen nucleus in a single collision (this is due to the kinematics of the collision) although the average fractional loss is approximately 50 percent per collision. However the neutron can not only "collide" with a nucleus, but can also be absorbed by it and not be available for completion of the reactor cycle. Consequently we need not only low-mass material, but also "inert" materials against neutron absorption. For this reason deuterium, D, (^{2}H) is often used since the probability of neutron capture in this nucleus is much less than in ^{1}H, although deuterium is more difficult to obtain and not as efficient at slowing down neutrons because of its greater mass. Typical materials used for slowing down neutrons are H_2O, D_2O, C, and Be. The process of slowing down neutrons is called *moderation,* and the material that does this in a reactor is called the *moderator.*

10.4 STATIC CYCLE PARAMETERS   We can follow a single neutron through a full fission cycle of neutron production and moderation to find the number of neutrons available at the end of the cycle to start the next one (see Fig. 10-4). When a neutron strikes a fuel nucleus and causes a fission, a total of ν neutrons are produced in the fission process; however not all absorbed neutrons produce fission, so the fission neutrons per incident neutron, η, is somewhat less than ν. These fast neutrons are capable of generating further fissions either in the ^{235}U or in whatever fraction of ^{238}U may be present (remember that natural uranium is a combination of both uranium isotopes). The fast neutron fission gain, ϵ, is not very great, typically 1.1 or less. Some of these fast neutrons can escape from the reactor, which reduces the number available to carry on the cycle. The remainder is slowed down in the moderator. During this process there will be a number absorbed in the moderator, and we will be left with $\epsilon \, \eta \, p$ thermal neutrons per initial neutron, where p is the nonabsorption probability for the moderation process, that is, the probability that the neutron will be slowed down to thermal energies in the moderator without having been absorbed, and therefore removed from the cycle. After the neutrons have been thermalized they may leak out of the reactor, or be absorbed in the reactor material before reinitiating the fission cycle. The probability of nonabsorption of the neutrons after thermalization but before entering the next cycle, is described by the term f. The terms p and f are determined by the properties of the reactor materials; the fast and slow neutron leakage probabilities are governed by the geometry of the fuel and reactor. We can temporarily ignore the escape problem by assuming an infinitely large reactor, and define a neutron cycle gain, which tells us whether a self-sustaining reaction can occur for the particular

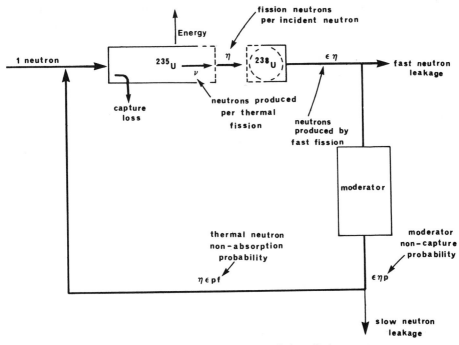

FIGURE 10-4 Schematic diagram of basic neutron cycle in a fission reactor.

choice of fuel and structural material for an adequately large reactor geometry to ensure negligible leakage losses. This factor is called k_∞

$$k_\infty = \eta \epsilon p f$$

If this number equals or exceeds unity it is possible to have a self-sustaining fission reaction for some size of reactor; if it is less than one it is not possible to continue the fission process at a constant rate without an additional supply of neutrons. In principle it should be possible to construct a self-sustaining reactor using thermal neutrons with ^{235}U, since at thermal energies $\nu = 2.41$ and $\eta = 2.1$. By a careful choice of materials and fuel arrangement it should be possible to have a practical (i.e., non-infinite size) self-sustaining reactor. For a finite dimension reactor the leakage factor must also be taken into account. When this is done the neutron cycle gain is given by the factor k;

$$k = k_\infty \times \text{geometric loss factor.}$$

Neutron losses due to these various factors can be quite small in optimized reactor designs. Approximately 0.1 neutron per cycle can be expected to be absorbed by fission products in the fuel, another 0.1 is lost by absorption in the structural materials (assuming the use of D_2O) and as few as 0.1 neutrons can be lost to control rod absorption and leakage from the reactor. Consequently k could take on any value up to as high as $\sim (\eta - 0.3)$, which for ^{235}U would be 1.8. If the value of k is less than unity for any reactor then the neutrons generated by a fission will not produce enough subsequent fissions to "replace" themselves so that the fission process will die out, and the reactor is described as *subcritical*. If $k = 1$ then there is exactly one neutron returned from any fission to start the next cycle, the fission rate will be constant and the reactor is described as *critical*. If $k > 1$ then more neutrons will be available to produce fission at the end of a cycle than were present at the start, and the reaction rate will increase indefinitely; the reactor is described as *supercritical*. Of these three conditions, the one we need for steady energy generation is $k = 1$.

10.5 CONTROL OF CRITICALITY   Once having designed k_∞ to ensure that criticality can be obtained for some nonzero leakage rate, it is necessary to estimate the likelihood of neutron leakage for the particular reactor assembly. This is largely controlled by the reactor dimensions. To see this we will take a homogeneous mixture of fuel and moderator, put it into the simplest possible geometric arrangement, a sphere of radius R. Since k is essentially a neutron use-to-loss ratio, and we know that the probability of use (fission) is proportional to the number of fuel nuclei present and that the probability of loss is proportional to the surface area available for leaking out, the value of k must be proportional to the ratio of these two values, hence

$$k \propto \frac{\text{Vol}}{\text{area}} = \frac{const\ R^3}{R^2} = \text{const} \times R$$

This is only a rough approximation, however, and does not properly treat attenuation of neutron flux. The actual variation of k with reactor dimension is linear only over small ranges of R, as shown in Fig. 10-5; we know it must approach the value k_∞ for

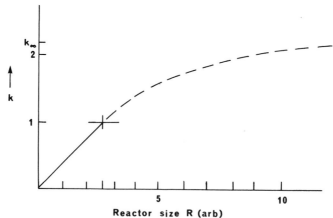

FIGURE 10-5 Approximate variation of k with reactor size.

very large dimensions. However we can see that there is a direct relation between reactor size and criticality.

10.6 CYCLE TIME AND THE IMPORTANCE OF BEING SUPERCRITICAL It is clear that by proper design and choice of dimension a reactor can be obtained that will operate at a constant power output. However to obtain a finite power level from the reactor we must be able to increase the power level (at least initially from zero) to some desired value and therefore we must allow k to become greater than 1 while the power level is being increased. What will happen to the reaction rate if we let k be greater than 1? What are the "dynamics" of our reactor?

To answer this question we must know the cycle time of the reactor; that is, how long between "generations" of neutrons. In estimating the cycle time we will assume that the time required for the fission process to occur is very short compared to the time required to thermalize the neutrons and to strike another fuel nucleus. In the following way we can estimate very crudely the time required to slow down. The probability of a neutron hitting a nucleus as it passes through one atomic layer of material is approximately the ratio of the nuclear cross-sectional area to the atomic cross-sectional area. This gives us

$$\text{collision probability per atomic layer} \approx \frac{\pi\, r_n^2}{\pi\, r_a^2} \cong \frac{(10 \times 10^{-15})^2}{(\sim 10^{-10})^2}$$

$$\cong 10^{-8}$$

Consequently we can expect that it will pass through $\sim 10^8$ atom layers before a neutron is likely to strike a nucleus and lose energy, a distance of the order of 1 cm; the fast neutron is initially going at an average velocity of approximately 10^9 cm/s so that a collision will occur in a time on the order of $\sim 10^{-9}$ seconds. If we estimate that half the neutron's energy is lost in each collision, then the number of collisions, n, required to reach an energy of the order of 0.02 eV from an energy of ~ 1 MeV should be given by the relation

$$(1/2)^n = \frac{0.02 \text{ eV}}{1 \text{ MeV}}$$

This gives a value of $n = 25$. How long will it take for this number of collisions to occur? Since we estimate that half the neutron's energy is lost in each collision, the velocity is reduced by a factor of $\sqrt{2}$ and therefore the time between collisions increases by $\sqrt{2}$ each time. We can set up the following formula for the total time taken to slow down to thermal velocities

$$t = t_0 + \sqrt{2}\, t_0 + (\sqrt{2})^2 t_0 + (\sqrt{2})^3\, t_0 \ldots \ldots + (\sqrt{2})^n t_0$$

where t_0 is 10^{-9} s. This can be solved fairly simply in an approximate way:

$$t = t_0(1 + x + x^2 \ldots \ldots x^n)$$

$$= t_0\, x^n \left(\frac{1}{x^n} + \frac{1}{x^{n-1}} \ldots \ldots + 1 \right)$$

$$\cong \frac{t_0\, x^n}{\left(1 - \dfrac{1}{x}\right)}$$

$$\cong \frac{10^{-9} \times (\sqrt{2})^{25}}{\left(1 - \dfrac{1}{\sqrt{2}}\right)} = 2 \times 10^{-5}\ \text{s.}$$

This is obviously a rough estimate, but it shows that the cycle time is short on a time scale of seconds. Typical values of slowing down times are given in Table 10-2. The values are in surprisingly good agreement with our rough calculation.

Given a typical value for the reactor cycle time, it is possible to look at what happens to the power level as a function of time while the reactor is supercritical. Rather than calculate the power, we will just work out the change in the number of neutrons available for fission as the cycles progress. The change in the number of neutrons in one cycle is given by

$$\frac{\Delta N}{\Delta t} = \frac{(k - 1)}{T_{\text{cyc}}} N$$

This can be put in the form of a differential equation

$$\frac{dN}{N} = \frac{(k - 1)dt}{T_{\text{cyc}}}$$

which has the solution

$$N(t) = N_0 e^{(k - 1)t/T_{\text{cyc}}}$$

where N_0 is the initial number of neutrons.

TABLE 10-2   Characteristics of Frequently Used Moderators

Moderator	Slowing down time (s)	Average number of collisions to thermalize
C	1.5×10^{-4}	114
H_2O	5.6×10^{-6}	19
D_2O	4.3×10^{-5}	35
Be	5.7×10^{-5}	86
fast fission cycle time		10^{-7} s

The rate of change of available neutrons, and therefore the change of reactor power level, is an exponential in time, controlled by the amount of supercriticality and the cycle time. To get a feeling for this let us take an extreme case; let $k = 2$ and $T_{cyc} = 10^{-7}$ s, and ask what happens in 1 second.

$$\frac{N}{N_0} = e\left(\frac{(2 - 1) \times 1}{10^{-7}}\right) = e^{10^7}$$

What occurs is described more readily as an explosion than a controlled increase in the power level! If the power level of a reactor is to increase in anything like a controllable fashion, the value of k cannot be far different from unity. How different? Let us see what k would have to be to allow an increase by a factor of e in the number of neutrons in a reasonable time (say 10 s).

$$\frac{N}{N_0} = e^1 = e\frac{(k - 1) \times 10}{10^7} = e^{(k - 1)10^8}$$

$$(k - 1) = 10^{-8}$$

$$k = 1.00000001!$$

This, among other things, would require an accuracy of the value of k for the reactor of something better than one part in 10^8! Although this is an extreme case for the cycle time it is clear that any safe reactor design requires much more than slide-rule accuracy. In fact one would be tempted from these results to conclude that it is not possible to build a usable reactor at all since it would not be possible to safely control its operating conditions.

10.7 DELAYED NEUTRONS AND REACTOR CONTROL   In fact if the previously discussed factors were the only way to control fission reactors, real reactor design and technology would be far more difficult than they are, if not impossible. However there is an additional property of the fission process which puts reactor control within the range of present-day capabilities; that is the presence of delayed neutrons.

Of the products formed in the fission process there are approximately six decay chains that behave in the manner shown in Fig. 10-6. One of the possible β decay modes of ^{87}Br goes to a highly excited state of ^{87}Kr; it is so highly excited that it is unbound to neutron emission. This excited state immediately neutron-decays to ^{86}Kr, which is stable. The neutron is emitted only after the decay of the ^{87}Br, which has a characteristic half life of 56 s. Total generation of neutrons by this delayed emission process in the various decay chains comprises approximately 0.65 percent of all fission neutrons from ^{235}U. Together the delayed groups have a characteristic production time of approximately 0.1 s. If it can be arranged that the reactor response is determined by the number of these

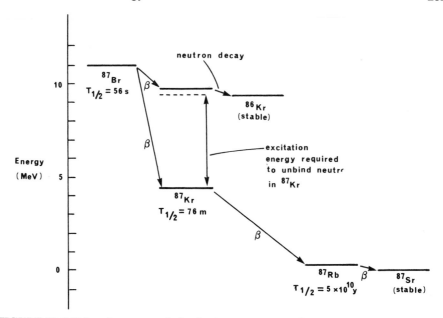

FIGURE 10-6 Delayed neutron emission in the decay chain of ^{87}Br. The absolute value of the energy scale is drawn relative to the ground state of ^{87}Sr.

delayed neutrons, it becomes possible to allow increases of the power level by letting k become greater than 1 in a controllable fashion. By designing the reactor to operate in the region $1 < k < 1.0065$, the excess of neutrons per cycle beyond that necessary to maintain a constant level will be due *entirely* to delayed neutrons, and only when they have been generated will there be a net increase in the fission rate. The characteristic cycle time for the rate of increase consequently will be controlled by the delayed neutrons, rather than the prompt neutrons, and the exponential increase will have a characteristic time of greater than $0.1/0.0065 \cong 12$ seconds.

The relatively long cycle time for delayed neutrons makes it possible to change the criticality back to unity quickly enough to retain control of the reactor. Quite clearly it is an important feature of reactors that they are designed so that k can never exceed 1.0065 and become uncontrollable. At the value $k = 1.0065$ a reactor using ^{235}U is said to be *prompt critical*.

Having seen that there is a range of k values that will allow a reactor to be operated in a controllable way, we must now look into the question of how to actually set k. The first control of the value of k is, as we discussed before, the geometry of the reactor itself. Having decided on a detailed arrangement and composition of fuel, moderator, structural elements, and so on, the reactor geometry must be arranged so that the maximum achievable k is less than 1.0065; this ensures that the reactor can never go prompt critical, and start increasing output power with a prompt cycle time. Without this limitation the design would not be intrinsically safe.

With the geometry set, the value of k can then be modified by introducing controlled amounts of neutron-absorbing materials into the reactor. This will have the general effect of reducing k, and by varying the amount of neutron absorber present (normally inserted in rods of some form, called control rods), k can be varied from the design maximum to almost zero. The materials used for the control rods or neutron absorbers are ones that have been found to have very high neutron absorption cross sections, such as B, Cd, or In.

In reactors that use a liquid moderator, such H_2O or D_2O, it is also possible to modify k by changing the amount of moderator present. If there is not enough moderator to thermalize all the neutrons produced, then their effectiveness at producing fissions is inhibited strongly, and consequently k is reduced. This procedure is not used to carry out the normal power level control and modification, but can be employed if a rapid reactor shut-down is required.

10.8 THE PHYSICAL REACTOR   The basic layout of a nuclear power reactor can now be set up. Primarily it consists of fuel (we have discussed only ^{235}U, but it could be any of the fissile elements) surrounded by moderator to optimize fission efficiency, and absorber to control the power level. Since the basic purpose of the reactor is to generate energy there must be some form of heat extraction. The type of coolant varies with reactor design, ranging from air, H_2O, D_2O, and organic liquids to inert gases and liquid metals. The competing requirements include low cost, low neutron absorption cross section, and high heat capacity. In addition, there must be a shield (or shields) to attenuate the flux of neutrons and γ rays from the reactor so that tolerable radiation levels can be achieved for personnel operating the reactor, and also to absorb and contain the energy release from any anticipated possible thermal explosion which extreme conditions might produce. Up to about 10 percent of the heat produced in a reactor comes from release of energy by β decay of fission products (see Table 9-1). This gives reactor systems two characteristics unique to the fission energy source. First, adequate cooling must be provided for a long time after the reactor has been ''turned off.'' Second, fuel that has been removed from a reactor is a strong radioactive source and must be dealt with as a serious biological hazard in all subsequent handling and storage. Although the biological effects of nuclear radiations is outside the scope of this book, anyone seriously interested in nuclear power should become familiar with this important topic. A very good introduction to the subject of radiation effects can be found in chapters 9 to 11 of Lamarsh (see Further Reading).

There are several variations of reactor design in operation. Many reactors use light water. The water may be allowed to boil in the reactor itself (Fig. 10-7) and produce steam which is carried directly to a turbine for electricity production. This system, called a boiling water reactor (BWR) has the construction advantage of low pressures which allow lighter and more compact construction (a separate steam generator is not required). Thermal efficiency can be improved by pressurizing the water so that higher

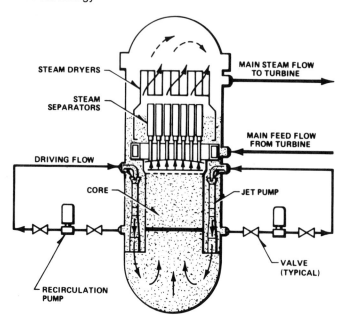

FIGURE 10-7 Boiling water reactor (BWR). Cooling water is forced through the reactor core by the jet pumps and steam is separated from the water at the top of the reactor vessel. Illustration courtesy of General Electric Co.

water temperatures can be reached without boiling. This kind of a reactor, called a pressurized water reactor (PWR), is shown in Fig. 10-8. It requires stronger construction to withstand safely the necessary high pressures, and also a separate steam generating apparatus. The higher thermal efficiency of pressurized water coolant is obtained at the expense of greater material costs and the potential liability to onset of boiling which impedes cooling ability, in case of accidental loss of pressurization. Because light water (H_2O) is a modest absorber of neutrons, light water reactors must use fuel enriched in its ^{235}U content (^{235}U is only 0.7% of natural uranium, the remainder being ^{238}U which at thermal energies only absorbs neutrons).

The effectiveness of neutrons at producing fission reactions can be increased by replacing the hydrogen in the moderator and coolant by deuterium, that is, heavy water. Heavy water reactors are able to use unenriched uranium as a fuel. An operating heavy water, natural uranium-fueled reactor system is shown in Fig. 10-9. In this type of reactor the use of a cheaper fuel is offset by a more expensive moderator and coolant, since D_2O is only present in minute quantities in normal water and must be concentrated to at least 99 percent D_2O for use in reactors.

In order to make more efficient use of the reactor heat energy various high temperature gas reactors have been developed. One such system is shown in Fig. 10-10. These typically use a carbon moderator (since it can withstand extremely high temperatures) and various cooling gases, over a range of temperatures. The earliest gas-cooled

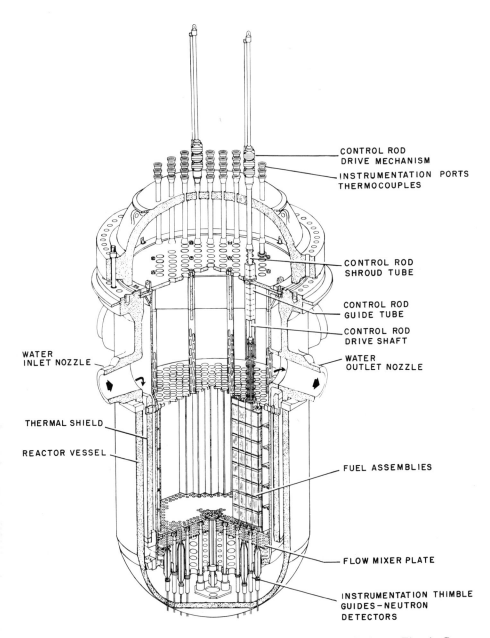

FIGURE 10-8 Pressurized water reactor (PWR). Courtesy of Westinghouse Electric Corporation.

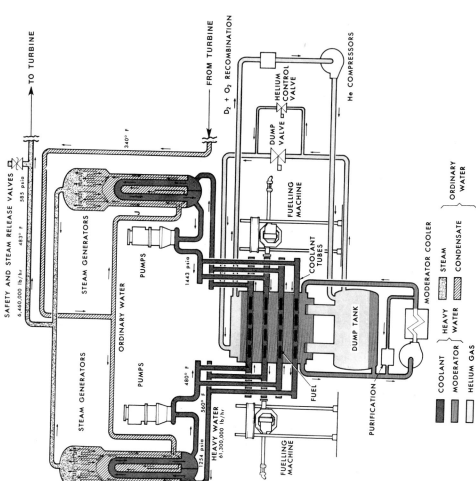

FIGURE 10-9 Simplified station flow diagram of heavy water cooled and moderated reactor. The reactor can be shut down by forcing moderator into dump tank so that fission neutrons are not slowed down. Fuel can be inserted and removed while the reactor is under power. Illustration

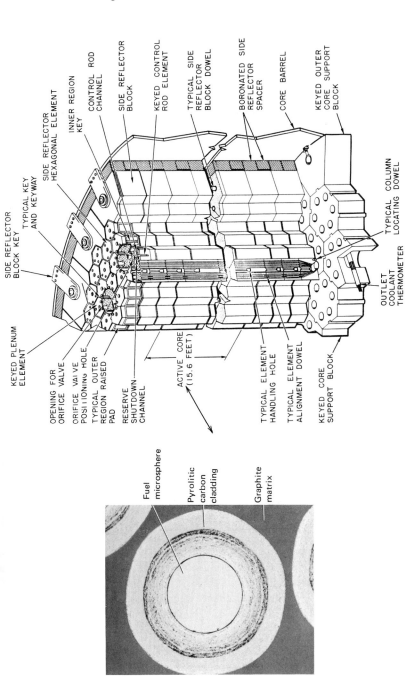

FIGURE 10-10 High temperature gas cooled reactor core assembly and expanded view of fuel arrangement. Blocks of graphite surrounding the core serve to reflect neutrons back into the core and the boronated blocks at the reactor edge minimize neutrons outside the reactor. Fuel microsphere by W. V. Goedell, *Neuclear Science and Engineering*, Vol. 20, 1964. Illustration of reactor and fuel microspheres, courtesy General Atomic Company.

reactors used CO_2, largely for economy, but as higher operating temperatures come into use ($\sim 1400°C$), the chemical stability and negligible neutron capture cross section of ^4He will make it a more desirable coolant for high temperature gas reactors.

10.9 FUEL BREEDING   In addition to producing fission, neutrons can be used to transmute nonfissile elements (ones that cannot undergo fission with thermal neutrons) into fissile ones. Typical of such reaction are those produced by $n + {}^{238}U$ and $n + {}^{232}Th$;

and

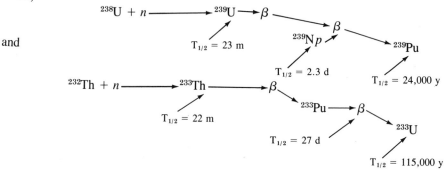

The end-products of these processes are both fissile materials; in fact all the elements listed in the fuel table as artificial can be produced by neutron bombardment and subsequent β decay.

The availability of ^{239}Pu alone as an additional fuel would mean at least a hundredfold increase (the $^{238}U/^{235}U$ abundance ratio) in the amount of fissile material. Taken together, the artificial fuels provide a major increase in the reactor fuel available and also an increased range of reactor operation parameters which might lead to an improved fuel economy. These fuels can be generated within the reactor itself since, in principle, the fission neutron gain is well in excess of unity required to keep the reactor running at a constant rate. Since the basic ^{235}U neutron gain factor, η, is 2.1, there could be in principle up to 1.1 neutrons available to breed fuel for each uranium fuel atom used; it should be possible to produce more fuel than is burned up. A reactor set up to do this is called a *breeder reactor*.

The process of breeding fuel is not quite as simple as it might at first appear. The elements to be made fissile (called fertile materials) cannot be subjected to too large a flux of neutrons, since the probability of absorbing a second neutron before the β-decay chain is complete would become sufficiently high to reduce the yield of fissile material. This together with the desirability of separating the produced fuel makes complex fuel handling procedures necessary.

In order to gain the maximum breeding yield per fission cycle it is clear that breeder reactors should be designed to have the highest possible (or practicable) value of η and minimum neutron losses. For this reason much effort has been put into the development of a cycle using ^{239}Pu, which has an η of 2.99 at a neutron energy of 2

MeV. The system designed to make use of this high value of η is called a *fast breeder reactor,* and the basic feature is the absence of a moderator, which ensures that a minimum neutron energy loss occurs before the fission neutrons strike a fuel atom. Because of this, fast breeder reactors can be much more compact. In order to separate material that is being used as fuel and material that is being used to produce fuel, the fuel is highly enriched in fissile material (^{235}U or ^{239}Pu or a combination) with the result that the thermal energy generated is much more concentrated than in thermal (i.e., slow neutron) reactors. The high power density and the need to minimize neutron moderation makes the use of a liquid coolant, such as sodium or potassium, necessary for such reactors, although the technical problems involved in their safe operation are much greater than for other forms of coolant. A prototype design for a fast breeder reactor is shown in Fig. 10-11.

The higher energy neutron spectrum of a fast reactor leads to a shorter cycle time ($\sim 10^{-7}$ s) and also has an effect on the fraction of delayed neutrons that are generated (Table 10.3). In the case of ^{239}Pu this fraction is reduced by about a factor of 3 from the normal ^{235}U thermal cycle. This means that the design criteria for the ^{239}Pu fast breeder cycle have to be much more strictly controlled. In terms of stringent design demands, both for heat extraction and criticality control, the fast reactor presents a much greater engineering challenge than the thermal reactor.

10.10 FUSION POWER   We have seen that it is possible to extract energy from the nucleus by fusion and by fission; however, although this possibility has been known for several decades, only fission power has become a practical energy source. Why is this?

Unlike the case for fission, where the reaction can be initiated by an (uncharged) neutron, there is at present no known way to produce fusion without the use of a reaction between two charged particles. This means that the particles must be given sufficient kinetic energy to overcome the Coulomb barrier before the fusion reaction can take place. For the lowest-mass particle that can cause a fusion reaction, $^2_1D + ^2_1D \rightarrow ^3_2He + n + 3.27$ MeV , the threshold energy is 8.6 keV (Fig. 10-12). Although this threshold energy is small compared to the energy released, the efficient production and use of charged particles with keV energies is not presently possible. Energetic charged particle beams can be produced readily and even efficiently by existing particle accelerators. However, when these particles "strike" target atoms the most likely thing to happen (compare the reaction cross sections in Fig. 10-12 to the physical cross section of any atom) is that energy will be transferred to the electrons of the target atom from the projectile via the Coulomb interaction. In this way the projectile very quickly loses its kinetic energy and most probably drops below the threshold energy before a fusion reaction can happen. Consequently, if fusion reactions are to occur with high efficiency, it is necessary to give the "target" atoms comparable kinetic energies to the "projectile" atoms so that kinetic energy is not lost over the great number of collisions required to ensure a fusion reaction. The simplest way to do this is to *heat* a mixture of "target" and "projectile" atoms to a temperature where

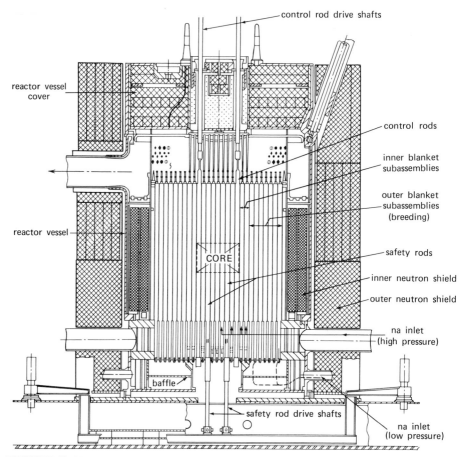

FIGURE 10-11 A prototype fast breeder reactor. The fuel consists of either ^{235}U or ^{239}Pu and is cooled by liquid sodium. A blanket of ^{238}U surrounds the fuel to provide the breeding material. From *Reactor Handbook,* 2nd ed., vol. IV, edited by S. McLain and J. H. Martens, Copyright © 1964 by John Wiley & Sons, Inc. Printed by permission of John Wiley & Sons, Inc.

TABLE 10-3 Delayed Neutron Fraction for Fast and Thermal Fission*

Element	Fast Fission	Thermal Fission
^{232}Th	0.022	—
^{233}U	0.0027	0.0026
^{235}U	0.0065	0.0064
^{238}U	0.015	—
^{239}Pu	0.0021	0.0021
^{240}Pu	0.0026	—

*From A.R. Foster and R.L. Wright, *Basic Nuclear Engineering,* 3rd ed, 1977, Allyn & Bacon

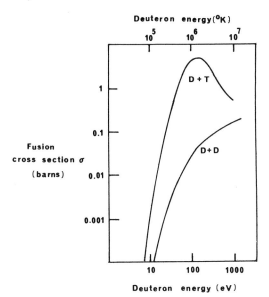

FIGURE 10-12 Energy dependence of deuteron induced fusion probability for deuterium, (D + D) and tritium (D + T) targets. The fusion probability is given as an effective cross section in units of barns (10^{-24} cm²).

a significant number of atoms have a kinetic energy greater than the reaction threshold energy, that is, $kT \geqslant E_{\text{thresh}}$. The D + D reaction threshold energy of 8.6 keV corresponds to 10^5 K, which is an extremely high temperature. The reaction $^2_1\text{D} + ^3_1\text{H} \rightarrow ^4_2\text{He} + n + 17.6$ MeV has a somewhat lower threshold energy, 4 keV (4.6×10^4 K) but this is still a very high temperature.

In addition to the very high ''ignition'' temperature required, the reaction kinematics show that a large part of the excess energy is carried off by the neutron. In the case of $^2\text{D} + ^3\text{H}$ reaction, the neutron takes away 14 MeV of the total excess energy. Since the neutrons are not charged, this energy can be converted to a practical form, heat, by a nuclear reaction. For this purpose the reaction

$$^6_3\text{Li} + n \rightarrow ^7_3\text{Li} \rightarrow ^3_1\text{H} + ^4_2\text{He}$$

can be used to absorb the neutron energy and convert it to motion of charged particles which can then be converted to thermal energy. It also has the desirable feature of regenerating the tritium (^3_1H) required to initiate the fusion reaction.

To date no device has been developed that can produce more fusion power than the power required to initiate fusion. However steady work, and progress, has been taking place for many years. At the moment there are two main methods being pursued to attain the desired net energy gain condition; these are plasma confinement and heating, and laser implosion heating.

The plasma confinement arrangement works by placing a hot net neutral gas of electrons and positive ions (plasma)—most likely a D-T mixture—inside a magnetic field region that does not allow the plasma to escape. An arrangement that appears to be quite effective is a torus (doughnut) as shown in Fig. 10-13. This has a magnetic field set up in such a way that the magnetic force $F = q\,\mathbf{v} \times \mathbf{B}$ forces the plasma away from the vessel walls and into a trajectory that follows the central ring of the torus. If a very large current is made to flow in this path, the current itself will provide the appropriate shape of magnetic force to compress the plasma into the center of the vessel.

Plasma heating can be done in several ways; the most straightforward way is to use the plasma current itself as the secondary circuit of a transformer to increase the plasma velocity and therefore its temperature. However it is necessary to provide an adequate power input to the plasma because, although the plasma is not in physical contact with the container walls, the accelerations caused by Coulomb collisions result in an energy loss from the plasma by electromagnetic radiation. Some methods of more rapid heating are presently being considered; the use of high-powered lasers at relatively long wavelengths in order to interact mainly with the outer electrons of the plasma, which then heat the main plasma mixture. Another method is to inject a beam of high-energy neutral particles into the plasma. The beam, being uncharged, will not be affected by the confining magnetic field and therefore can be injected readily into the plasma. Collisions with the plasma particles will quickly ionize the beam and allow it to interact with the plasma, thereby heating it.

A more recent development than magnetic plasma confinement has become possible with the advent of very high power lasers. It is possible to induce fusion by sudden irradiation of a solid pellet composed of a mixture of deuterium and tritium (D-T). The laser pulse causes both strong heating and a shock-wave compression of the material which can achieve extremely high densities and temperatures while the pellet implosion is going on. Compared to magnetic plasma confinement techniques, this

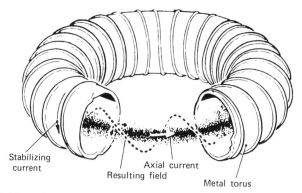

Stabilizing current

Axial current

Resulting field

Metal torus

FIGURE 10-13 Magnetic confinement of a plasma in a toroidal system. From The Potential of Nuclear Fusion by R. S. Pease, in *Contemporary Physics*, 18, March 1977.

method attains much higher particle densities (hence greater D-T collision probability, resulting in fusion) but these conditions can only last for a very short time. A tentative design of such a system is shown in Fig. 10-14.

The laser optical components are likely to be affected by neutrons from the fusion reaction. The neutrons will produce optical absorption centers in the lenses, which could extract unacceptably large amounts of energy from the laser beams (melt the lenses). Such potential difficulties recommend an alternative method being considered in which the laser beams are replaced with beams of energetic ions that do not use optical lens materials. (In this system the difficulty lies in the ability to produce adequately high power ion beams and sources.)

No proposed fusion technique at present is able to achieve the combined requirements of high particle density and long confinement times that are necessary to obtain a net energy gain from the fusion reaction. Approximately an order of magnitude improvement in both particle density and confinement time is required before the energy break-even point is achieved. The development of a working fusion reactor will occupy

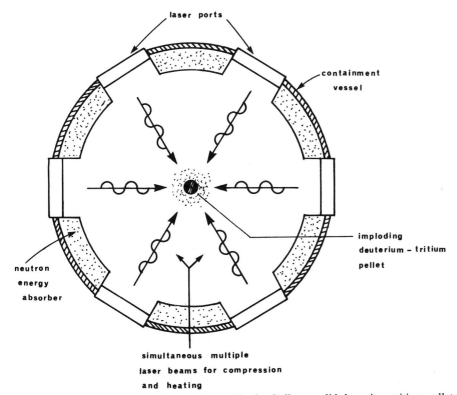

FIGURE 10-14 Laser-induced fusion achieved by imploding a solid deuterium-tritium pellet. The kinetic energy of the resulting neutrons would be absorbed by a wall of lithium lining the vacuum vessel.

a large fraction of the innovative engineering effort for several years, quite probably decades, to come.

10.11 FISSION-FUSION SYMBIOSIS   Fission and fusion devices can be regarded as having complementary characteristics which argue for some form of linked operation for their mutual benefit. The fission process is well understood and controlled with present technology but the most effective utilization of neutrons for production of supplementary fuel in addition to maintaining energy output still requires considerable effort and ingenuity. On the other hand, it is likely that neutrons will be produced in fusion reactors long before the impediments to achievement of net energy output are overcome. In fact these fusion neutrons have a value beyond their kinetic energy, in that they can be used to convert fertile materials such as ^{232}Th or ^{238}U into ^{233}U and ^{239}Pu for use in fission reactors. In this way the neutrons become capable of producing an amount of energy far in excess of what they could in a fusion cycle alone.

Tentative designs of fusion-fission systems have centered around plasma fusion devices, likely because of the constant neutron flux levels present from steady-state devices. The basic modification, as indicated in Fig. 10-15, is to replace part of the neutron-absorbing blanket around the fusion plasma with either a fission fuel breeding blanket (subcritical), or a reactor core configuration. The system will also most likely

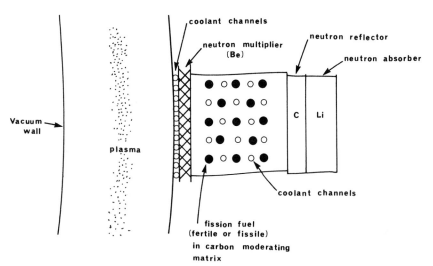

FIGURE 10-15 Schematic diagram of subcritical fission energy convertor for neutrons produced by fusion reactor. Neutrons from the fusion reaction can produce additional neutrons when they strike the beryllium surrounding the vacuum vessel. These neutrons can be used to either produce fission power or breed fissile fuel.

include a material, such as beryllium, which readily emits additional neutrons when bombarded with energetic neutrons, so that the number of neutrons available for fuel breeding (or inducing fission) is maximized.

The detailed design of a combined fission-fusion system will require evaluation of questions such as what the optimum running efficiency for both fission and fusion should be, whether they should be linked together physically in one hybrid device or be run separately with one providing otherwise unavailable fuel and the other ''subsidizing'' an otherwise energetically uneconomical process. Such choices will depend heavily on the detailed nature of improved understanding that will be achieved by further research on fusion of machines.

FURTHER READING H. L. Davis, ed., Special Issue: Magnetically Confined Fusion, *Physics Today* 32, May 1979.

A. R. Foster and R. L. Wright, *Basic Nuclear Engineering,* 3rd ed., Allyn and Bacon, 1977.

J. H. Fremlin, *Applications of Nuclear Physics,* The English Universities Press, 1964.

D. I. Garber and R. R. Kinsey, *Neutron Cross Sections,* vol. II, Curves, Brookhaven National Laboratory Report *325,* 3rd ed. 1976.

J. R. Lamarsh, *Introduction to Nuclear Engineering,* Addison Wesley, 1975.

L. M. Lidsky, Fission-Fusion Systems: Hybrid, Symbiotic and Augean, *Nuclear Fusion 15* (1975) 151.

S. McLain and J. H. Martens, eds., *Reactor Handbook,* 2nd ed., vol. IV, Wiley, 1964.

A. M. Weinberg and E. P. Wigner, *The Physical Theory of Neutron Chain Reactions,* University of Chicago Press, 1958.

J. Weisman, ed., *Elements of Nuclear Reactor Design,* Elsevier, 1977.

P. F. Zweifel, *Reactor Physics,* McGraw-Hill, 1973.

PROBLEMS 1. The *activity* of a sample is its rate of radioactive decay. Show that the activity follows the same time behavior as the total number of atoms does.

2. Calculate the rate of energy release from 1 kg of plutonium due to its natural activity. Use the decay information in the Table of Nuclides. Calculate the rate of energy release from 1 kg of ^{60}Co.

3. A reaction cross section can be interpreted as a simple geometrical cross sectional area. Estimate the probability per atom of a neutron causing a fission if the fission cross section is 1000 barns (see Appendix 1).

4. Using the neutron-induced fission probabilities shown in Fig. 10-1 estimate the relative amount of fuel required for a critical mass (sufficient material so that $k = 1$)

in two reactors that are the same except that in one the average neutron energy is 0.03 eV and the other one it is 10 keV.

5. Since the reactor temperature and average neutron energy are related, the temperature at which a reactor is operated will influence the value of the fission probability per atom, and therefore the amount of fuel required to run the reactor at a fixed power level. Estimate the fractional change in the fission probability per atom if a reactor's temperature is changed from 500°C to 600°C.

6. The estimate of the relation of k to reactor size given in the text was clearly good only for small changes in k (at best) since reduction of neutron number (attenuation) was not considered. How should the neutron escape probability vary with reactor size for a single neutron source at the center of a spherical neutron absorber (reactor) of radius R? How would this result have to be modified to account for a homogeneous neutron source, which is the situation in a real reactor? (Merely set up the equations, do not attempt to solve this problem.)

7. What is the maximum fraction of its initial kinetic energy that can be lost by a projectile in an elastic collision if it strikes an object whose mass is (a) equal to the projectile mass, (b) ten times the projectile mass, (c) two hundred times the projectile mass? You can use conservation of energy and momentum to obtain an answer.

8. Neutrons have an interaction probability per atom, or cross section, for a boron target of 759 barns for a neutron energy of 0.03 eV. What thickness of boron would be required to reduce the intensity of a beam of neutrons by a millionfold (see Appendix 1)?

9. What is the exponential characteristic time for the rate of neutron increase in a reactor if $k = 1.005$, with a delayed neutron production time of 0.1 s? What is it if $k = 1.0065$ and $T_{cycle} = 10^{-4}$ s?

10. Attenuation coefficients of 1 MeV γ rays in water and lead are shown in Fig. 9-8. If you assume the interaction cross section of 1 MeV *neutrons* in these two materials is roughly equal to the geometrical size of the element, compare the relative effectiveness of these two materials at interacting with 1 MeV neutrons and γ rays.

11. The spectrum of fission neutrons is produced by very highly excited nuclear matter (compared to its characteristic binding energies), suggesting that a classical description of the neutron energetics might be appropriate. How well does the observed neutron spectrum confirm this expectation? How do you expect the induced fission neutron energy spectrum to be related to (a) incident neutron energy, and (b) fissioning element?

12. Using the known breakdown of forms of energy release following fission, estimate the fraction of power from a reactor that is due to β decay. What do you expect to happen to this power when the reactor shuts down? What is actually observed? Why?

13. You have been asked to look into ways to detect leakage of fission products from fuel rods, both while they are in the reactor and after they have been removed for storage in cooling water. Discuss the possibility of detecting delayed neutrons for this purpose in the two cases.

14. From the known fusion probability (cross section) and the geometrical cross section of an atom, estimate the probability of a fusion occurring when a deuteron passes by a tritium target atom. If the deuteron loses ~ 5 eV per atom it passes, what is the probability of induced fusion per incident deuteron if it has an initial energy of 1 MeV? Assuming no energy losses in accelerating the deuteron to this energy, what would the efficiency of such a fusion process be? How reasonable is the estimate for energy loss per atom?

15. Estimate what the fusion rate per cm^3 of a 10 atmosphere D-T plasma would be if the temperature of the plasma is 10^6 K. Assume all particles are traveling at the average velocity and that the fusion cross section is 1 b. Use the approximations discussed in problems 2-6 and 2-10.

17. The $^6Li + p \rightarrow {}^3He + {}^4He$ reaction is being considered by some as a possible fusion energy source. Calculate the energy release (the reaction Q) and compare it with other fusion reaction energy gains (e.g. problems 9-23 and 9-24). What do you think the advantages and disadvantages of this reaction would be?

18. A basic problem of practical fusion energy sources is bringing two nuclei close enough and long enough for fusion to occur. Diatomic molecules can be thought of as nuclei permanently kept within a certain distance of each other. Discuss the possibility of inducing fusion by "chemical" means.

APPENDIX ONE

The quantitative description of the probability that a beam of particles (atoms, nucleons, photons) interact with particles in a target is normally given in terms of an interaction *cross sections*. Although cross sections can be interpreted easily in geometrical terms, it is quantitatively more useful to use the following operational definition. If there is a beam consisting of n particles per unit volume all traveling with a velocity v, then the quantity

$$I = nv$$

which is called the beam *intensity*, gives the number of particles that will pass through a surface one unit in area perpendicular to the beam in one second. If a target of area A is placed in the beam, then

$$IA = nvA$$

particles will strike the target in one second. If the target is very thin, so that any particle in the beam strikes at most one atom in the target, and no atoms are shielded from the beam by atoms in front of them, it is observed experimentally that the number of interactions ("collisions") per second is proportional to the beam intensity I, the target atom density N, and the target area A, and thickness dx. This proportionality is given in the relation

Number of interactions per second $= \sigma I N A dx$

by the symbol σ, which is called the *cross section*. It clearly has units of area. The numerical value of σ depends on the type and energy of the particles in the beam and the kind of target atoms. Once determined, it completely fixes the interaction rate through the above equation.

There is a more physical interpretation that can be given to the concept of cross section. As stated above, IA particles interact with the target per second. The number of atoms in the target is $NAdx$, so that the number of particles per second that interact with an individual target atom is given by the total number of interaction per second divided by the total number of target atoms, which is $I\sigma$. Therefore the probability that an *individual* particle will interact with an individual target atom must be given by

$$\frac{I\sigma}{IA} = \frac{\sigma}{A}$$

If we picture an interaction as a physical collision in order to generate a simple physical picture, the above relation indicates that σ can be thought of as just the fractional area of the target occluded by a single target atom. That is, it is an *effective cross-sectional area* that the individual atom presents to the particles in the beam for the particular kind of interaction being considered.

The concept of cross section has been presented above in terms of particles interacting with atoms. The same type of behavior is found to occur for atomic projectiles interacting with targets of atoms, nucleon projectiles interacting with targets of nuclei, and photons interacting with targets of either atoms or nuclei. All of these kinds of interaction rates therefore can be calculated using the concept of cross section. Characteristic cross sections are *generally* of the same order of magnitude as the physical size of the entities involved both for atom-atom and nucleon-nucleus interactions; that is, atomic cross sections are likely to be of the order of magnitude of squared ångstroms (10^{-16} cm^2) and nuclear cross sections are likely to be of the order of barns (10^{-24} cm^2). Photon interactions are not amenable to such a simple geometric picture, and are much smaller than particle cross sections, both in the atomic and nuclear case.

APPENDIX TWO

It is often useful to be able to estimate what the effective "size," or cross section, and overall behavior is when energetic charged particles such as electrons or atoms scatter off other atoms. These processes are controlled by the Coulomb interaction, and can be treated classically in most cases of practical interest.

The trajectory of a charged particle past an infinitely massive repulsive center is illustrated in Fig. A-1, and is fixed by two facts. First, total energy, shared between kinetic energy and Coulomb potential energy, is conserved. This gives us the relation

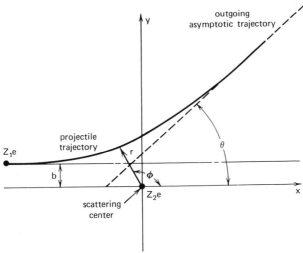

FIGURE A-1 Trajectory of particle with charge Z_1e scattering from an infinitely massive scattering center at the origin with charge Z_2e.

$$\frac{m}{2}\left[\left(\frac{dr}{dt}\right)^2 + \left(\frac{rd\phi}{dt}\right)^2\right] + \frac{kZ_1Z_2e^2}{r} = \frac{mv_0^2}{2}$$

where m and Z_1e are the mass and charge of the projectile, v_0 is its initial velocity, Z_2e is the scattering center's charge (its mass is infinite) and dr/dt and $rd\phi/dt$ are the projectile's radial and tangential velocity, respectively. Second, since the Coulomb force is central, there can be no torque so that angular momentum must be conserved. This gives us

$$mr^2\frac{d\phi}{dt} = mv_0b$$

where the quantity b, called the *impact parameter*, is the closest approach the projectile would have in the absence of a force. The initial angular momentum about the scattering center is given by mv_0b. The latter relation gives us

$$\frac{d\phi}{dt} = \frac{v_0b}{r^2}$$

so that after expressing dr/dt as $dr/d\phi\ d\phi/dt$, the kinetic energy relation becomes

$$\frac{m}{2}\left[\left(\frac{dr}{d\phi}\right)^2 \frac{v_0^2b^2}{r^4} + \frac{v_0^2b^2}{r^2}\right] + \frac{kZ_1Z_2e^2}{r} = \frac{mv_0^2}{2}$$

This can be simplified by dividing both sides by $mv_0^2/2$;

$$\left(\frac{dr}{d\phi}\right)^2 \frac{b^2}{r^4} + \frac{b^2}{r^2} + \frac{kZ_1Z_2e^2}{mv_0^2r/2} = 1$$

The third term on the left-hand side can be made more tractable by making the substitution

$$\frac{mv_0^2}{2} \equiv KE_0$$

and

$$\frac{kZ_1Z_2e^2}{b} \equiv PE_b$$

this leaves

$$\left(\frac{dr}{d\phi}\right)^2 \frac{b^2}{r^4} + \frac{b^2}{r^2} + \frac{PE_b}{KE_0}\frac{b}{r} = 1$$

The solution of this simple differential equation is given by

$$\phi = -\arcsin\left[\frac{2KE_0b/PE_br + 1}{\sqrt{1 + 4(KE_0/PE_b)^2}}\right] + \phi_0$$

where the constant ϕ_0 is fixed by the fact that at $r = \infty$, $\phi = \pi$, so that

$$\phi_0 = \pi + \arcsin \frac{1}{\sqrt{1 + 4(KE_0/PE_b)^2}}$$

This gives, after some manipulation,

$$\frac{1}{r} = \frac{\sqrt{1 + 4(KE_0/PE_b)^2}}{2bKE_0/PE_b} \left[\sin\left(\phi - \arcsin \frac{1}{\sqrt{1 + 4(KE_0/PE_b)^2}} \right) - \frac{1}{\sqrt{1 + 4(KE/PE_b)^2}} \right]$$

As $r \to \infty$ the value of ϕ approaches the scattering angle θ. This occurs when the bracketed quantity in the above equation is zero. Therefore

$$\sin\left(\theta - \arcsin \frac{1}{\sqrt{1 + 4(KE_0/PE_b)^2}} \right) = \frac{1}{\sqrt{1 + 4(KE_0/PE_b)^2}}$$

or

$$\theta = 2 \arcsin \frac{1}{\sqrt{1 + 4(KE_0/PE_b)^2}}$$

From trigonometry we recognize that this gives

$$\cot \frac{\theta}{2} = 2 \frac{KE_0}{PE_b} = \frac{mv_0^2 b}{kZ_1Z_2e^2}$$

which relates the scattering angle θ to the initial conditions KE_0 and b and Coulomb properties Z_1 and Z_2 of the target-projectile pair.

Having determined the relation that fixes the scattering angle, we can now obtain an estimate of a total scattering probability, or cross section. Because the $1/r^2$ behavior that characterizes the Coulomb force acts for all distances, it is necessary to choose some minimum deflection angle, θ_{min}, below which we will assume that no significant interaction has taken place. The choice of θ_{min} has a corresponding maximum impact parameter b_{max}, beyond which it may be assumed that no interaction occurs. Therefore all projectiles that start towards the target within a circle of radius b_{max} around the impact parameter $b = 0$ will be scattered by the target; all others will not. The area of this circle defines the scattering cross section, so that

$$\sigma(\theta > \theta_{min}) = \pi b^2$$

or, in terms of initial parameters

$$\sigma(\theta > \theta_{min}) = \pi \cot^2 \frac{\theta_{min}}{2} \left(\frac{kZ_1Z_2e^2}{mv_0^2} \right)^2$$

These scattering relations can be applied in a straightforward manner when the scattering center and projectiles are structureless charged objects. When one or both of the objects have significant charge structure, such as atoms, the process can be treated by means of *effective charges*. These can be determined quite simply by means of the Bohr model, or, more accurately, from the radial charge distribution discussed in sec. 4-2.

APPENDIX THREE

Stable nuclides are indicated by an asterisk (*). The atomic masses listed are from A. H. Wapstra and N. B. Gove, Nuclear Data Tables 9 (1971), 267. For each A, only those columns where a change occurs are listed in the mass values.

A	El	Atomic Mass (amu)	A	El	Atomic Mass (amu)	A	El	Atomic Mass (amu)	A	El	Atomic Mass (amu)
1	n	1.00866		Be	1692	12	Be	12.02678	16	C	16.01470
	H*	782		B	2999		B	1435		N	6.00610
							C*	0000		O*	5.99491
2	H*	2.01410	8	He	8.03397		N	1861		F	6.01147
				Li	2248						
3	H	3.01604		Be	0530	13	Be	13.03835	17	C	17.01886
	He*	602		B	2460		B	1778		N	7.00845
							C*	0335		O*	6.99913
			9	Li	9.02680		N	0573		F	7.00209
4	H	4.02783		Be*	1218		O	2480		Ne	7.01769
	He*	0260		B	1332						
	Li	2697		C	3103						
						14	B	14.02602	18	N	18.01425
5	H	5.03627	10	Li	10.03794		C	0324		O*	7.99915
	He	1222		Be	1353		N*	0307		F	8.00093
	Li	1254		B*	1293		O	0859		Ne	8.00571
				C	1685						
6	He	6.01889									
	Li*	512	11	Li	11.04649	15	B	15.03157	19	N	19.01755
	Be	972		Be	2166		C	1059		O	9.00357
				B*	0930		N*	0010		F*	8.99840
7	He	7.02803		C	1143		O	0307		Ne	9.00188
	Li*	1600		N	2732		F	1896		Na	9.01393

A	El	Atomic Mass (amu)	A	El	Atomic Mass (amu)	A	El	Atomic Mass (amu)	A	El	Atomic Mass (amu)
20	O	20.00407		Si*	7692	38	S	37.97116		Ti*	5262
	F	19.99998		P	9232		Cl	6800		V	6020
	Ne*	19.99244					Ar*	6273			
	Na	20.00735	29	Al	28.98044		K	6909	47	K	46.96167
	Mg	20.01880		Si*	7649		Ca	7635		Ca	5454
				P	8180					Sc	5241
21	O	21.01146		S	9662	39	Cl	38.96800		Ti*	5176
	F	0.99995					Ar	6431		V	5490
	Ne*	0.99384	30	Al	29.98294		K*	6370		Cr	6296
	Na	0.99765		Si*	7377		Ca	7071			
	Mg	1.01171		P	7831				48	Ca*	47.95252
				S	8490	40	Cl	39.97044		Sc	5223
22	F	22.00303					Ar*	6238		Ti*	4794
	Ne*	1.99138	31	Si	30.97536		K	6400		V	5225
	Na	1.99443		P*	7376		Ca*	6259		Cr	5403
	Mg	1.99958		S	7960		Sc	7797			
	Al	2.01937		Cl	9227				49	Ca	48.95567
						41	Ar	40.96450		Sc	5002
23	Ne	22.99447	32	Si	31.97413		K*	182		Ti*	4787
	Na*	2.98977		P	7390		Ca	227		V	4851
	Mg	2.99412		S*	7207		Sc	925		Cr	5127
	Al	3.00726		Cl	8576					Mn	5951
						42	Ar	41.96305			
24	Ne	23.99361	33	P	32.97172		K	6240	50	Ca	49.95751
	Na	3.99096		S*	7145		Ca*	5862		Sc	5218
	Mg*	3.98504		Cl	7745		Sc	6553		Ti*	4478
	Al	3.99994		Ar	8991		Ti	7303		V	4716
	Si	4.01155								Cr*	4604
			34	P	33.97335	43	K	42.96072		Mn	5424
25	Na	24.98995		S*	6787		Ca*	5877			
	Mg*	4.98583		Cl	7376		Sc	6116	51	Sc	50.95359
	Al	4.99043		Ar	8025		Ti	6852		Ti	4660
	Si	5.00410								V*	4396
			35	S	34.96903	44	K	43.96156		Cr	4477
26	Na	25.99193		Cl*	6885		Ca*	5548		Mn	4821
	Mg*	8259		Ar	7525		Sc	5940			
	Al	8689		K	8792		Ti	5969	52	Ti	51.94689
	Si	9232								V	478
			36	S*	35.96707	45	K	44.96069		Cr*	051
27	Na	26.99293		Cl	6830		Ca	5619		Mn	556
	Mg	6.98434		Ar*	6754		Sc*	5591		Fe	811
	Al*	6.98154		K	8141		Ti	5813			
	Si	6.98670					V	6575	53	V	52.94432
	P	7.00023	37	S	36.97111					Cr*	065
				Cl*	6590	46	K	45.96196		Mn	129
28	Mg	27.98387		Ar	6677		Ca*	5368		Fe	531
	Al	8191		K	7337		Sc	5517			
				Ca	8580						

A	El	Atomic Mass (amu)	A	El	Atomic Mass (amu)	A	El	Atomic Mass (amu)	A	El	Atomic Mass (amu)
54	V	53.94640		Cu	3257	71	Zn	70.92771		Br	2114
	Cr*	3888		Zn	3439		Ga*	2470		Kr*	2040
	Mn	4035					Ge	2495			
	Fe*	3961	63	Co	62.93358		As	2711	79	Ge	78.92550
	Co	4846		Ni	2966		Se	3248		As	2089
				Cu*	2958					Se	1848
55	Cr	54.94082		Zn	3320	72	Zn	71.92686		Br*	1833
	Mn*	3804		Ga	3911		Ga	636		Kr	2008
	Fe	3829					Ge*	208		Rb	2386
	Co	4201	64	Co	63.93547		As	675			
				Ni*	2795		Se	740	80	As	79.92297
56	Cr	55.94067		Cu	2975					Se*	1652
	Mn	3890		Zn*	2914	73	Ga	72.92513		Br	1853
	Fe*	3493		Ga	3673		Ge*	2346		Kr*	1637
	Co	3983					As	2382		Rb	2260
	Ni	4212	65	Ni	64.93007		Se	2677			
				Cu*	2778		Br	3182	81	As	80.92208
57	Mn	56.93814		Zn	2923					Se	1799
	Fe*	539		Ga	3273	74	Ga	73.92708		Br*	1629
	Co	628		Ge	3950		Ge*	2117		Kr	1661
	Ni	977					As	2393		Rb	1904
			66	Ni	65.92908		Se*	2247		Sr	2311
58	Mn	57.93982		Cu	2886		Br	2999			
	Fe*	327		Zn*	2603		Kr	3332	82	Se*	81.91670
	Co	3575		Ga	3159					Br	679
	Ni*	3533		Ge	3385	75	Ga	74.92642		Kr*	348
	Cu	4453					Ge	2287		Rb	820
			67	Ni	66.93215		As*	2160		Sr	885
59	Fe	58.93486		Cu	2774		Se	2252			
	Co*	3318		Zn*	2713		Br	2576	83	Se	82.91901
	Ni	3434		Ga	2820		Kr	3123		Br	1517
	Cu	3949		Ge	3296					Kr*	1413
						76	Ge*	75.92140		Rb	1524
60	Fe	59.93404	68	Cu	67.92976		As	2239		Sr	1766
	Co	3381		Zn*	484		Se*	1921		Y	2249
	Ni*	3077		Ga	798		Br	2469			
	Cu	3735		Ge	839		Kr	2576	84	Se	83.91849
	Zn	4182								Br	1655
			69	Cu	68.92922	77	Ge	76.92361		Kr*	1150
61	Fe	60.93663		Zn	2655		As	2064		Rb	1438
	Co	245		Ga*	2557		Se*	1991		Sr*	1343
	Ni*	105		Ge	2796		Br	2137		Y	2089
	Cu	346		As	3223		Kr	2459			
	Zn	926	70	Zn*	69.92532				85	Br	84.91554
				Ga	2602	78	Ge	77.92295		Kr	1253
62	Co	61.93394		Ge*	2425		As	2189		Rb*	1179
	Ni*	2833		As	3093		Se*	1730		Sr	1294

A	El	Atomic Mass (amu)	A	El	Atomic Mass (amu)	A	El	Atomic Mass (amu)	A	El	Atomic Mass (amu)
	Y	1644		Mo	1176		Mo*	0602		Ag	1157
	Zr	2170		Tc	1778		Tc	0639		Cd	1469
							Ru	0763		In	2521
86	Br	85.91845	92	Rb	91.91946		Rh	1137			
	Kr*	1061		Sr	1098		Pd	1653	103	Mo	102.91358
	Rb	1117		Y	0892					Tc	0885
	Sr*	0927		Zr*	0503	98	Zr	97.91275		Ru	0633
	Y	1493		Nb	0719		Nb	1035		Rh*	0551
	Zr	1633		Mo*	0680		Mo*	0541		Pd	0610
				Tc	1534		Tc	0711		Ag	0898
87	Br	86.92034					Ru*	0528		Cd	1371
	Kr	1336	93	Rb	92.92157		Rh	1071		In	2079
	Rb	0918		Sr	1417		Pd	1265		Sn	2927
	Sr*	0889		Y	0955						
	Y	1091		Zr	0644	99	Zr	98.91588	104	Mo	103.91391
	Zr	1467		Nb*	0638		Nb	1105		Tc	1112
	Nb	2025		Mo	0680		Mo	0772		Ru*	0542
				Tc	1022		Tc	0625		Rh	0666
88	Kr	87.91444		Ru	1688		Ru*	0593		Pd*	0401
	Rb	1132					Rh	0814		Ag	0841
	Sr*	0562	94	Sr	93.91547		Pd	1179		Cd	0981
	Y	0951		Y	1169		Ag	1828		In	1894
	Zr	1024		Zr*	0632					Sn	2387
	Nb	1797		Nb	0728	100	Zr	99.91725			
	Mo	2291		Mo*	0508		Nb	1392	105	Tc	104.91140
				Tc	0966		Mo*	0747		Ru	0775
89	Kr	88.91781		Ru	1135		Tc	0784		Rh	0569
	Rb	1228					Ru*	0421		Pd*	0508
	Sr	0746	95	Sr	94.91891		Rh	0811		Ag	0651
	Y*	0586		Y	1279		Pd	0854		Cd	0952
	Zr	0890		Zr	0803		Ag	1638		In	1500
	Nb	1306		Nb	0682					Sn	2209
	Mo	1947		Mo*	0583	101	Zr	100.92173		Sb	3218
				Tc	0766		Nb	1476			
90	Kr	89.91961		Ru	1041		Mo	1035	106	Tc	105.91431
	Rb	1487		Rh	1578		Tc	0732		Ru	0732
	Sr	0775					Ru*	0557		Rh	0728
	Y	0716	96	Y	95.91559		Rh	0617		Pd*	0348
	Zr*	0471		Zr*	0829		Pd	0830		Ag	0668
	Nb	1127		Nb	0809		Ag	1303		Cd*	0646
	Mo	1394		Mo*	0467		Cd	1897		In	1369
				Tc	0783					Sn	1724
91	Kr	90.92324		Ru*	0759	102	Nb	101.91820		Sb	2948
	Rb	1626		Rh	1451		Mo	1025			
	Sr	1016					Tc	0918	107	Ru	106.91013
	Y	0730	97	Y	96.91752		Ru*	0434		Rh	0675
	Zr*	0564		Zr	1096		Rh	0684		Pd	0512
	Nb	0699		Nb	0810		Pd*	0560		Ag*	0509

A	El	Atomic Mass (amu)	A	El	Atomic Mass (amu)	A	El	Atomic Mass (amu)	A	El	Atomic Mass (amu)
	Cd	0661		Sb	1213		Sb	0558		Xe*	0612
	In	1036		Te	1664		Te	0590		Cs	1245
	Sn	1605					I	1245			
	Sb	2474	113	Ag	112.90656		Xe	1599	125	Sn	124.90779
	Te	3554		Cd	0440					Sb	0524
				In*	0408	119	Cd	118.90959		Te*	0442
108	Ru	107.91014		Sn	0518		In	0584		I	0458
	Rh	0872		Sb	0937		Sn*	0331		Xe	0645
	Pd*	0389		Te	1571		Sb	0393		Cs	0974
	Ag	0595		I	2344		Te	0640		Ba	1457
	Cd*	0418					I	0984			
	In	0972	114	Ag	113.90873		Xe	1519	126	Sn	125.90766
	Sn	1197		Cd*	0336					Sb	0732
	Sb	2249		In	0490	120	In	119.90822		Te*	0331
	Te	2991		Sn*	0277		Sn*	0220		I	0562
				Sb	0888		Sb	0508		Xe*	0427
109	Rh	108.90864		Te	1178		Te*	0402		Cs	0965
	Pd	0595		I	2112		I	0971		Ba	1158
	Ag*	0475					Xe	1208			
	Cd	0495	115	Ag	114.90885				127	Sn	126.91035
	In	0711		Cd	0543	121	In	120.90787		Sb	0691
	Sn	1120		In	0387		Sn	0423		Te	0521
	Sb	1839		Sn*	0335		Sb*	0382		I*	0447
	Te	2762		Sb	0660		Te	0490		Xe	0518
				Te	1148		I	0744		Cs	0743
110	Rh	109.91096		I	1738		Xe	1151		Ba	1119
	Pd*	0516		Xe	2541					La	1656
	Ag	0611				122	In	121.91064			
	Cd*	0301	116	Ag	115.91152		Sn*	0345	128	Sn	127.91046
	In	0723		Cd*	0476		Sb	0517		Sb	0907
	Sn	0786		In	0526		Te*	0305		Te*	0446
	Sb	1667		Sn*	0174		I	0750		I	0581
	Te	2257		Sb	0658		Xe	0868		Xe*	0353
				Te	0825					Cs	0772
111	Pd	110.90765		I	1609	123	In	122.91045		Ba	0848
	Ag	0528		Xe	2092		Sn	0573		La	1578
	Cd*	0418					Sb*	0422			
	In	0507	117	Cd	116.90723		Te	0428	129	Sb	128.90918
	Sn	0776		In	0453		I	0557		Te	0659
	Sb	1302		Sn*	0296		Xe	0844		I	0498
	Te	2097		Sb	0484		Cs	1295		Xe*	0478
				Te	0859					Cs	0597
112	Pd	111.90738		I	1322	124	In	123.91323		Ba	0859
	Ag	0706		Xe	1973		Sn*	0528		La	1289
	Cd*	0276					Sb	0594			
	In	0554	118	Cd	117.90692		Te*	0283	130	Sb	129.91160
	Sn*	0483		In	0612		I	0622		Te*	0623
				Sn*	0161					I	0672

A	El	Atomic Mass (amu)	A	El	Atomic Mass (amu)	A	El	Atomic Mass (amu)	A	El	Atomic Mass (amu)
	Xe*	0351	136	I	135.91474		Pr*	0769		Eu	1725
	Cs	0675		Xe*	0722		Nd	0963		Gd	1855
	Ba*	0628		Cs	0729		Pm	1364		Tb	2724
	La	1240		Ba*	0455		Sm	1836			
131	Sb	130.91187		La	0764				147	Ce	146.92246
	Te	0854		Ce*	0718	142	Cs	141.92370		Pr	1902
	I	0612		Pr	1276		Ba	1651		Nd	1612
	Xe*	0508		Nd	1540		La	1414		Pm	1516
	Cs	0546					Ce	0930		Sm*	1492
	Ba	0690	137	I	136.91754		Pr	1008		Eu	1681
	La	1008		Xe	1174		Nd*	0776		Gd	1931
	Ce	1470		Cs	0707		Pm	1294		Tb	2425
				Ba*	0581		Sm	1514			
132	Sb	131.91455		La	0635				148	Ce	147.92412
	Te	0854		Ce	0764	143	Ba	142.92054		Pr	2219
	I	0800		Pr	1059		La	1604		Nd*	1692
	Xe*	0415		Nd	1489		Ce	1240		Pm	1749
	Cs	0641					Pr	1085		Sm	1485
	Ba*	0504	138	Xe	137.91404		Nd*	0985		Eu	1817
	La	1010		Cs	1103		Pm	1100		Gd	1819
	Ce	1160		Ba*	0523		Sm	1473		Tb	2422
				La	0716		Eu	2011		Dy	2723
133	Sb	132.91519		Ce*	0602				149	Pr	148.92337
	Te	1100		Pr	1079	144	Ba	143.92292		Nd	2015
	I	0783		Nd	1219		La	1959		Pm	1836
	Xe	0589					Ce	1368		Sm*	1721
	Cs*	0543	139	Xe	138.91844		Pr	1334		Eu	1803
	Ba	0598		Cs	1328		Nd	1012		Gd	1940
	La	0803		Ba	0883		Pm	1267		Tb	2337
	Ce	1157		La*	0640		Sm*	1207		Dy	2756
				Ce	0669		Eu	1886		Ho	3390
134	Te	133.91136		Pr	0896		Gd	2284	150	Pr	149.92629
	I	0985		Nd	1197	145	La	144.92173		Nd*	2092
	Xe*	0539		Pm	1680		Ce	1722		Pm	2106
	Cs	0670					Pr	1454		Sm*	1730
	Ba*	0449	140	Xe	139.92137		Nd*	1261		Eu	1978
	La	0847		Cs	1676		Pm	1279		Gd	1870
	Ce	0901		Ba	1063		Sm	1347		Tb	2371
	Pr	1567		La	0952		Eu	1639		Dy	2582
				Ce*	0548		Gd	2177		Ho	3344
135	Te	134.91650		Pr	0912	146	La	145.92546		Er	3784
	I	1006		Nd	0963		Ce	1869	151	Nd	150.92388
	Xe	0713		Pm	1596		Pr	1753		Pm	2124
	Cs	0589	141	Cs	140.91963		Nd*	1315		Sm	1996
	Ba*	0567		Ba	1415		Pm	1473		Eu*	1988
	La	0679		La	1093		Sm	1310			
	Ce	0926		Ce	0832						
	Pr	1310									

A	El	Atomic Mass (amu)	A	El	Atomic Mass (amu)	A	El	Atomic Mass (amu)	A	El	Atomic Mass (amu)
	Gd	2038		Yb	4587	160	Eu	159.93179		Hf	4519
	Tb	2318		Lu	5417		Gd*	2707		Ta	5442
	Dy	2640					Tb	2720			
	Ho	3183					Dy*	2523	165	Dy	164.93174
	Er	3752	156	Sm	155.92554		Ho	2836		Ho*	3035
	Tm	4536		Eu	2477		Er	2922		Er	3075
				Gd*	2214		Tm	3545		Tm	3243
152	Nd	151.92471		Tb	2461		Yb	3846		Yb	3539
	Pm	2340		Dy	2432		Lu	4683		Lu	3968
	Sm*	1975		Ho	2980		Hf	5145		Hf	4516
	Eu	2177		Er	3163					Ta	5182
	Gd	1981		Tm	3914	161	Gd	160.92969		W	5944
	Tb	2391		Yb	4302		Tb	2759			
	Dy	2479		Lu	5284		Dy*	2697	166	Dy	165.93283
	Ho	3165					Ho	2785		Ho	3232
	Er	3513	157	Eu	156.92543		Er	3004		Er*	3032
	Tm	4426		Gd*	2397		Tm	3383		Tm	3358
				Tb	2404		Yb	3866		Yb	3386
153	Pm	152.92405		Dy	2550		Lu	4456		Lu	3977
	Sm	2212		Ho	2819		Hf	5111		Hf	4266
	Eu*	2126		Er	3237					Ta	5104
	Gd	2151		Tm	3753	162	Gd	161.93098		W	5641
	Tb	2345		Yb	4288		Tb	2948			
	Dy	2583		Lu	5029		Dy*	2683	167	Ho	166.93312
	Ho	3040		Hf	5808		Ho	2916		Er*	3207
	Er	3532					Er*	2882		Tm	3288
	Tm	4214	158	Eu	157.92781		Tm	3387		Yb	3498
	Yb	4922		Gd*	2412		Yb	3634		Lu	3828
				Tb	2545		Lu	4396		Hf	4289
154	Pm	153.92652		Dy*	2444		Hf	4826		Ta	4869
	Sm*	2222		Ho	2871					W	5556
	Eu	2301		Er	3032	163	Tb	162.93057		Re	6383
	Gd*	2089		Tm	3741		Dy*	2877			
	Tb	2454		Yb	4063		Ho	2878	168	Ho	167.93538
	Dy	2447		Lu	4943		Er	3007		Er*	3240
	Ho	3065		Hf	5472		Tm	3267		Tm	3423
	Er	3301					Yb	3664		Yb*	3392
	Tm	4151	159	Eu	158.92923		Lu	4180		Lu	3861
	Yb	4632		Gd	2640		Hf	4781		Hf	4075
				Tb*	2538		Ta	5543		Ta	4827
155	Sm	154.92464		Dy	2577					W	5278
	Eu	2290		Ho	2760	164	Tb	163.93280		Re	6244
	Gd*	2263		Er	3093		Dy*	2921			
	Tb	2354		Tm	3544		Ho	3026	169	Ho	168.93693
	Dy	2579		Yb	4081		Er*	2923		Er	3462
	Ho	2934		Lu	4746		Tm	3348		Tm*	3424
	Er	3343		Hf	5412		Yb	3467		Yb	3521
	Tm	3955					Lu	4154		Lu	3765

A El	Atomic Mass (amu)	A El	Atomic Mass (amu)	A El	Atomic Mass (amu)	A El	Atomic Mass (amu)
Hf	4127	Ir	6855	178 Yb	177.94689	Ir	5819
Ta	4621	Pt	7762	Lu	4614	Pt	6179
W	5244			Hf*	4372	Au	6984
Re	5974	174 Tm	173.94217	Ta	4578	Hg	7567
Os	6822	Yb*	3888	W	4587		
		Lu	4035	Re	5087	183 Hf	182.95360
170 Ho	169.93946	Hf	4014	Os	5388	Ta	5141
Er*	3549	Ta	4443	Ir	6139	W*	5026
Tm	3583	W	4647	Pt	6665	Re	5086
Yb*	3479	Re	5334	Au	7650	Os	5344
Lu	3848	Os	5807			Ir	5709
Hf	3977	Ir	6740	179 Lu	178.94729	Pt	6192
Ta	4621	Pt	7423	Hf*	4584	Au	6806
W	4997			Ta	4596	Hg	7486
Re	5877	175 Tm	174.94387	W	4715		
Os	6500	Yb	4129	Re	5004	184 Ta	183.95422
		Lu*	4079	Os	5412	W*	5097
171 Er	170.93805	Hf	4144	Ir	5959	Re	5270
Tm	3645	Ta	4381	Pt	6603	Os*	5259
Yb*	3635	W	4703	Au	7387	Ir	5766
Lu	3786	Re	5165	Hg	8268	Pt	6034
Hf	4065	Os	5777			Au	6764
Ta	4462	Ir	6485	180 Lu	179.95012	Hg	7252
W	4999	Pt	7319	Hf*	4657		
Re	5632			Ta	4756	185 Ta	184.95557
Os	6405	176 Tm	175.94703	W*	4670	W	5346
Ir	7292	Yb*	4258	Re	5077	Re*	5300
		Lu	4270	Os	5281	Os	5409
172 Er	171.93936	Hf*	4142	Ir	5936	Ir	5678
Tm	3841	Ta	4476	Pt	6385	Pt	6086
Yb*	3640	W	4583	Au	7276	Au	6619
Lu	3909	Re	5174	Hg	7928	Hg	7217
Hf	3952	Os	5560				
Ta	4489	Ir	6395	181 Hf	180.94912	186 Ta	185.95859
W	4789	Pt	7011	Ta*	4802	W*	5440
Re	5573			W	4822	Re	5503
Os	6120	177 Yb	176.94527	Re	5016	Os	5388
Ir	7143	Lu	4377	Os	5341	Ir	5799
		Hf*	4324	Ir	5803	Pt	5974
173 Er	172.94266	Ta	4448	Pt	6366	Au	6618
Tm	3965	W	4663	Au	7057	Hg	6998
Yb*	3823	Re	5050	Hg	7840		
Lu	3897	Os	5555			187 W	186.95719
Hf	4069	Ir	6179	182 Hf	181.95072	Re	5579
Ta	4381	Pt	6921	Ta	5018	Os*	5578
W	4810	Au	7786	W*	4824	Ir	5740
Re	5368			Re	5131	Pt	6051
Os	6055			Os	5250	Au	6484

A	El	Atomic Mass (amu)	A	El	Atomic Mass (amu)	A	El	Atomic Mass (amu)	A	El	Atomic Mass (amu)
	Hg	7004		Pb	7687		Pb	7354		Bi	7761
	Tl	7691		Bi	8611		Bi	7945		Po	8080
							Po	8630		At	8880
188	W	187.95852	193	Os	192.96417		At	9413		Rn	9405
	Re	5815		Ir*	6296						
	Os*	5587		Pt	6303	198	Ir	197.97262	203	Au	202.97556
	Ir	5891		Au	6410		Pt*	6789		Hg	2.97287
	Pt	5949		Hg	6662		Au	6822		Tl*	2.97234
	Au	6519		Tl	7112		Hg*	6674		Pb	2.97340
	Hg	6831		Pb	7681		Tl	7046		Bi	2.97683
	Tl	7647		Bi	8390		Pb	7207		Po	2.98128
				Po	9206		Bi	7948		At	2.98720
189	W	188.96195					Po	8418		Rn	2.99377
	Re	5926	194	Os	193.96522		At	9349		Fr	3.00176
	Os*	5818		Ir	6511						
	Ir	5872		Pt*	6271	199	Pt	198.97057	204	Au	203.97833
	Pt	6076		Au	6540		Au	6876		Hg*	3.97349
	Au	6398		Hg	6546		Hg*	6827		Tl	3.97386
	Hg	6849		Tl	7136		Tl	6978		Pb	3.97304
	Tl	7461		Pb	7480		Pb	7278		Bi	3.97777
	Pb	8180		Bi	8350		Bi	7794		Po	3.98020
				Po	8950		Po	8405		At	3.98727
190	Re	189.96190					At	9135		Rn	3.99177
	Os*	5848	195	Os	194.96795					Fr	4.00120
	Ir	6068		Ir	6580	200	Pt	199.97143			
	Pt	5996		Pt*	6480		Au	7068	205	Hg	204.97607
	Au	6469		Au	6504		Hg*	6832		Tl*	7443
	Hg	6684		Hg	6655		Tl	7095		Pb	7448
	Tl	7435		Tl	6999		Pb	7171		Bi	7738
	Pb	7940		Pb	7493		Bi	7815		Po	8099
				Bi	8147		Po	8215		At	8604
191	Re	190.96300		Po	8895		At	9096		Rn	9183
	Os	6096					Rn	9676		Fr	9917
	Ir*	6063	196	Ir	195.96837						
	Pt	6170		Pt*	6496	201	Pt	200.97477	206	Hg	205.97752
	Au	6374		Au	6655		Au	7191		Tl	5.97612
	Hg	6729		Hg*	6582		Hg*	7030		Pb*	5.97447
	Tl	7265		Tl	7055		Tl	7074		Bi	5.97839
	Pb	7910		Pb	7312		Pb	7268		Po	5.98034
	Bi	8672		Bi	8128		Bi	7697		At	5.98645
				Po	8662		Po	8237		Rn	5.99025
192	Os*	191.96151		At	9644		At	8900		Fr	5.99884
	Ir	6264					Rn	9628		Ra	6.00461
	Pt*	6107	197	Ir	196.96950						
	Au	6485		Pt	6734	202	Au	201.97440	207	Tl	206.97744
	Hg	6582		Au*	6654		Hg*	7064		Pb*	6.97590
	Tl	7258		Hg	6699		Tl	7197		Bi	6.97848
				Tl	6957		Pb	7202		Po	6.98160

A El	Atomic Mass (amu)	A El	Atomic Mass (amu)	A El	Atomic Mass (amu)	A El	Atomic Mass (amu)
At	6.98573	Fr	1.99615	Ac	0931	Th	147
Rn	6.99060	Ra	1.99968	Th	1290	Pa	554
Fr	6.99722	Ac	2.00782				
Ra	7.00418			218 Po	218.00900	225 Ra	225.02363
		213 Pb	212.99664	At	0871	Ac	323
208 Tl	207.98201	Bi	2.99439	Rn	0561	Th	396
Pb*	7.97665	Po	2.99286	Fr	0752	Pa	612
Bi	7.97973	At	2.99293	Ra	0715		
Po	7.98125	Rn	2.99388	Ac	1153	226 Ra	226.02543
At	7.98654	Fr	2.99618	Th	1320	Ac	611
Rn	7.98953	Ra	3.00018			Th	491
Fr	7.99715	Ac	3.00657	219 At	219.01132	Pa	789
Ra	8.00218	Th	3.01312	Rn	0950		
				Fr	0924	227 Ra	227.02920
209 Tl	208.98536	214 Pb	213.99984	Ra	1008	Ac	2777
Pb	8.98109	Bi	3.99873	Ac	1243	Th	2772
Bi*	8.98040	Po	3.99521	Th	1543	Pa	2880
Po	8.98243	At	3.99634			U	3102
At	8.98617	Rn	3.99537	220 At	220.01526		
Rn	8.99021	Fr	3.99886	Rn	139	228 Ra	228.03109
Fr	8.99592	Ra	3.99997	Fr	232	Ac	3103
Ra	9.00210	Ac	4.00671	Ra	103	Th	2873
Ac	9.01008	Th	4.01154	Ac	475	Pa	3100
				Th	576	U	3138
210 Tl	209.99009	215 Bi	215.00185				
Pb	09.98419	Po	4.99944	221 Rn	221.01545	229 Ra	229.03511
Bi	09.98413	At	4.99865	Fr	425	Ac	296
Po	09.98288	Rn	4.99874	Ra	392	Th	178
At	09.98704	Fr	5.00035	Ac	559	Pa	209
Rn	09.98956	Ra	5.00273	Th	818	U	351
Fr	09.99622	Ac	5.00641			Np	625
Ra	10.00053	Th	5.01160	222 Rn	222.01760		
Ac	10.00961			Fr	1756	230 Ra	230.03713
		216 Bi	216.00643	Ra	1539	Ac	627
211 Pb	210.98876	Po	0191	Ac	1778	Th	315
Bi	0.98729	At	0242	Th	1848	Pa	455
Po	0.98665	Rn	0028	Pa	2348	U	395
At	0.98750	Fr	0319			Np	779
Rn	0.99061	Ra	0349	223 Fr	223.01976		
Fr	0.99549	Ac	0863	Ra	1852	231 Ac	231.03857
Ra	1.00077	Th	1095	Ac	1913	Th	631
Ac	1.00801			Th	2068	Pa	590
		217 Po	217.00641	Pa	2399	U	629
212 Pb	211.99190	At	0472			Np	827
Bi	1.99128	Rn	0393	224 Fr	224.02333		
Po	1.98887	Fr	0463	Ra	021	232 Ac	232.04205
At	1.99074	Ra	0632	Ac	171	Th	3807
Rn	1.99071						

A	El	Atomic Mass (amu)	A	El	Atomic Mass (amu)	A	El	Atomic Mass (amu)	A	El	Atomic Mass (amu)
	Pa	3859	235	Pa	235.04545	238	Pa	238.05507		Am	5553
	U	3715		U	394		U	5081		Cm	5552
	Np	4005		Np	407		Np	5097		Bk	5982
	Pu	4118		Pu	529		Pu	4958		Cf	6240
				Am	796		Am	5205			
							Cm	5305	241	Np	241.05833
233	Th	233.04160	236	Pa	236.04891		Bk	5824		Pu	5686
	Pa	4026		U	4558					Am	5684
	U	3965		Np	4663	239	U	239.05432		Cm	5767
	Np	4082		Pu	4605		Np	5295		Bk	6025
	Pu	4300		Am	4942		Pu	5217		Cf	6356
				Cm	5151		Am	5303			
							Cm	5486	242	Np	242.06177
234	Th	234.04363	237	Pa	237.05122		Bk	5829		Pu	5876
	Pa	335		U	4874					Am	5957
	U	097		Np	4819	240	U	240.05662		Cm	5885
	Np	291		Pu	4843		Np	5607		Bk	6208
	Pu	333		Am	5008		Pu	5382		Cf	6371
	Am	770		Cm	5280						

INDEX